Humanity's End

Life and Mind: Philosophical Issues in Biology and Psychology
Kim Sterelny and Robert A. Wilson, editors

Humanity's End

Why We Should Reject Radical Enhancement

Nicholas Agar

A Bradford Book
The MIT Press
Cambridge, Massachusetts
London, England

First MIT Press paperback edition, 2013

This book was set in Stone Sans and Stone Serif by Toppan Best-set Premedia Limited.

Library of Congress Cataloging-in-Publication Data

Agar, Nicholas.
Humanity's end : why we should reject radical enhancement / Nicholas Agar.
    p.   cm.—(Life and mind)
"A Bradford book."
Includes bibliographical references and index.
ISBN 978-0-262-01462-5 (hardcover : alk. paper)
ISBN 978-0-262-52517-6 (paperback : alk. paper) 1. Human evolution—Effect of technological innovations on. 2. Technological innovations—Social aspects.
I. Title.
GN281.A33   2010
303.48'3—dc22

2010003166

Contents

Acknowledgments

This book grew out of a short review article that Joyce Griffin and Greg Kaebnick asked me to write for the *Hastings Center Report*.[1] I'm grateful both for their prompting me to think seriously about radical enhancement and for their constructive critiques of my early efforts.

Many debts (all nonpecuniary) were incurred on the journey from a 3,000-word review to a 70,000-word book. My good friends Stuart Brock, Joseph Bulbulia, and Timothy Irwin formed a committee of readers whose skepticism prevented many errors and deepened discussion in many places—all for the price of a few Cab Savs and Sav Blancs. I should point out that their conclusion was pretty much "thanks for telling us about the millennial life spans and super-intelligence, now where can we get some." Score three to the advocates of radical enhancement. Mark Walker, one of radical enhancement's most astute defenders, passed a critical eye over drafts of all of the chapters. To forestall Mark's expulsion from the World Transhumanist Association I should say that he did his best to purge the book of its humanist biases. I'm grateful to Rob Wilson, an editor of the series in which this book appears. Rob provided valuable comments on the manuscript and invited me to present an earlier version of chapter 4 at the Human Kinds symposium at the 2009 Pacific Division of the American Philosophical Association in Vancouver. The other symposiasts offered helpful comments. My colleague Diana Burton did an excellent job helping to transform some disorganized early thoughts about radical enhancement into philosophical critiques. Big thanks go to Jan Agar who read and gave expert editorial advice on the entire manuscript.

My critical targets, Nick Bostrom, Aubrey de Grey, and James Hughes, were all generous with their time and advice. I'm sure Ray Kurzweil also would have been, had he not been too busy offering advice to the

incoming Obama administration and founding a tertiary education institu-
tion. It goes without saying that many differences remain—Hughes has
promised a "vigorous" review.

I owe thanks to Tom Stone who originally commissioned the project
for MIT Press and Philip Laughlin who very capably stepped in after Tom's
departure from the Press. Judy Feldmann's expert copyediting fixed many
broken sentences and tightened expression throughout.

Among others who offered insights along the way are Sondra Bacharach,
Ramon Das, David Eng, Matt Gers, Peter Hutchinson, Edwin Mares, Alice
Monro, Dan Turton, and David Wasserman.

Finally, I'm grateful to my wife, Laurianne, and kids, Alexei and Rafael,
who put up with me throughout. They were refreshing human reference
points through all my investigations of the various wacky and terrifying
forms that posthumanity might take.

1 What Is Radical Enhancement?

Suppose someone makes you the following offer: They will boost your intellect to such an extent that your cognitive achievements far exceed those of Einstein, Picasso, Mozart, or any of our familiar exemplars of genius. You'll have a huge range of new experiences much more marvelous than climbing Mt. Everest, being present at full orchestra performances of Beethoven's Ninth Symphony, or consuming peyote. You'll live for thousands of years. And the years you will gain won't have the diminished quality of those that modern medicine tends to provide. There'll be no need for oxygen bottles, Zimmer frames, bifocals, or any of the other standard accessories of extreme age in the early twenty-first century. The extra years will come with a guarantee of perfect health.

These are examples of what I will call *radical enhancement*. Radical enhancement involves improving significant human attributes and abilities to levels that greatly exceed what is currently possible for human beings. The radical enhancers who are this book's subjects propose to enhance intellects well beyond that of the genius physicist Albert Einstein, boost athletic prowess to levels that dramatically exceed that of the Jamaican sprint phenomenon Usain Bolt, and extend life spans well beyond the 122 years and 164 days achieved by the French super-centenarian Jeanne Calment.

Is the offer of radical enhancement one that should be taken seriously? If you received it by e-mail you might be tempted to file it in the folder reserved for unsolicited and ungrammatical offers of millions of dollars from the estates of people you've never met, and invitations to participate in multilevel marketing schemes. We will be exploring the ideas of a group of thinkers who make the offer in full earnestness. They propose to radically enhance our intellects and extend our life spans not by waving magic wands, but instead by administering a variety of technologies and thera-pies. These technologies and therapies are not available yet. But they could be soon. The writers of the *Star Trek* television series envisage our descen-dants two hundred years hence acquiring the ability to speedily travel to light-years-distant stars. If contemporary advocates of radical enhance-ment are right, then the individuals making those voyages will be more like the cerebral science officer Spock or the android Lieutenant Com-mander Data than the endearingly and annoyingly human Captain Kirk. It's also possible that we'll be alive to pilot the starships alongside them.

Prominent among advocates of radical enhancement is a social move-ment known as *transhumanism*. Nick Bostrom, the movement's foremost philosopher, defines transhumanism as an "intellectual and cultural move-ment that affirms the possibility and desirability of fundamentally improv-ing the human condition through applied reason, especially by developing and making widely available technologies to eliminate aging and to greatly enhance human intellectual, physical, and psychological capacities."[1] Transhumanists do not have a monopoly on the idea of radical enhance-ment, however. Some would-be radical enhancers are far too busy working out how to boost intellects and extend lives to bother affiliating themselves with an Internet-based social movement.

Fantastic offers tend to have catches, and radical enhancement is no exception to this rule. I'll argue that there's another side to the vision of millennial life spans and monumental intellects. Radical enhancement threatens to turn us into fundamentally different kinds of beings, so dif-ferent that we will no longer deserve to be called human. It will make us "posthuman." Although the benefits of radical enhancements of our minds and extensions of our lives may seem obvious—so obvious that they scarcely require defense—there is much that we stand to lose as we make the transition from human to posthuman. The aim of this book is to bring the costs of radical enhancement properly into focus. Some readers may find that its rewards are worth the price. But I suspect that many will not.

A Recent History of Radical Enhancement

Radical enhancement is not a new idea. Religion and myth are rich in accounts of humans seeking and undergoing radical enhancement. Those who transcribed myths or transmitted God's messages were alert to the possibility of a downside for such dramatic transformations. In ancient Greek myth, Zeus grants Tithonus the gift of eternal life. One thing he doesn't grant, however, is eternal *youth*. This leads, in some versions of the myth, to Tithonus' withering away to a cicada, a state in which (so the myth goes) he can be observed today. For Christians, the chief venue for radical enhancement is heaven. This is where the faithful experience an eternity of bliss in God's presence. According to many Christians, there's another, hotter location that offers a somewhat inferior version of eternal existence.

What's common to these stories is the view that one undergoes radical enhancement through the intercession of divine beings or by interacting with supernatural forces. The figures discussed in this book don't advocate prayer. They're offering DIY—do-it-yourself—radical enhancement. The idea that humans could be radically enhanced is not new; but the notion that it's something that we could arrange for ourselves certainly is.

Some reflections of the British biologist Julian Huxley make a convenient starting point for a recent history of DIY radical enhancement. Huxley was one of the twentieth century's greatest popularizers of science, a man driven by the belief that important choices should be informed by the relevant facts. Right at the top of Huxley's list of important choices were those concerning humanity's future.[2] He claimed to have noticed the possibility for a shift in our species' evolution that was both for the better and fundamental. The blind and uncaring forces of evolutionary change had given us conceptual thought and consciousness, gifts that empowered us to take charge of our own evolution. Huxley wrote, "[i]t is as if man had been suddenly appointed managing director of the biggest business of all, the business of evolution—appointed without being asked if he wanted it, and without proper warning and preparation. What is more, he can't refuse the job."[3] Huxley thought that the very best thing we could arrange for our species would be its transcendence, shifting human existence as it is has always been, "a wretched makeshift, rooted in ignorance," toward "a state of existence based on the illumination of knowledge and comprehension."[4]

A lot of things have happened since 1957, the year in which Huxley's call for transcendence was published, but a concerted, conscious redirection of human evolution doesn't seem to be among them. Indeed, humanity seems a good deal closer to extinction from greenhouse gases or weapons of mass destruction than it is to transcendence. What has changed between now and then that might make radical enhancement possible for us?

When Huxley first proposed it, radical enhancement was really little more than wishful thinking about the direction in which evolution might take us. Huxley's hope for transcendence stood in contrast to the more pessimistic view about humanity's evolutionary future presented by a sometime collaborator of his, the author and essayist H. G. Wells. Wells's novel *The Time Machine* depicted a humanity split into two descendant species. The ruling classes, accustomed to being waited on by their social inferiors, had given rise to the pretty but brainless and indolent Eloi. The working classes take a kind of delayed, evolutionary revenge by becoming the bestial Morlocks and periodically emerging from underground dwellings to capture and eat the Eloi. If you are accustomed to thinking of evolution as a process of continual improvement then you might have difficulty seeing how we could possibly evolve into the Eloi and the Morlocks. But evolution's concept of improvement differs from our own. One biological design is better than another in evolutionary terms if it conduces more efficiently to survival and reproduction. *Homo habilis* became the more intelligent *Homo erectus*, which in turn became the even more intelligent *Homo sapiens*, because, in the past, increases in intelligence have tended to boost the chances of surviving and having children. Wells imagines a future in which this correlation between increases in intelligence and greater success at surviving and reproducing no longer holds. If stupid, lazy people reliably have more children than intelligent industrious people, then the former are better adapted to their environments, and therefore "better" in evolutionary terms than the latter.

Contemporary advocates of radical enhancement want to use a variety of technologies to help us to avoid Wells's outcome. The really useful thing about the technologies that are this book's focus is that they can serve our ideals about improvement rather than those of evolution. Bostrom presents a concept of posthumanity that conceptually connects it with radical enhancement. He defines a posthuman as "a being that has at least one posthuman capacity," where a posthuman capacity is "a general central capacity greatly exceeding the maximum attainable by any current human being without recourse to new technological means."[5] According to Bostrom, general central capacities include, but are not limited to, "healthspan," which he understands as "the capacity to remain fully healthy, active, and productive, both mentally and physically"; cognition, which comprises intellectual capacities such as "memory, deductive and analogical reasoning, and attention, as well as special faculties such as the capacity to understand and appreciate music, humor, eroticism, narration, spirituality, mathematics"; and emotion, "the capacity to enjoy life and to respond with appropriate affect to life situations and other people."[6] Bostrom thinks that the desirability of improving our central capacities is obvious. We all want to live longer, be healthier, reason better, and feel happier. The only surprise is that we may not have to content ourselves with the small improvements offered by better diets, exercise programs, and bridge lessons. The right technologies can boost our mental powers and physical constitutions to levels far beyond those previously attained by humans.

The technologies that are the focus of this book can change us in the ways that petitioners for radical enhancement want, and can do so quickly. The dial on the time machine that takes Wells's hero to the Earth of the Eloi and Morlocks is set at AD 802,701; this fits with what Wells knew, and what we now know about evolution. It takes time for evolution to effect changes as large as those between us and the Eloi and Morlocks. This is good news for those who fear that we or our immediate descendants might suddenly become indolent or bestial dullards. But it's bad news for those aiming at transcendence. Huxley has some suggestions about how to speed up human evolution while ensuring that it takes us in the right direction. But we should not exaggerate the degree of this acceleration. Today's aspiring radical enhancers don't want to wait 800,000, or 700,000, or even 10,000 or 1,000 years for their vision of our future to be realized. They don't view extended life spans and superior intellects as gifts to be enjoyed by our great, great, great . . . grandchildren. They want these things for themselves—and if not for themselves, for their children. The technologies that we will examine in this book will not only be better than evolution at changing us in ways that we want, they'll also work much faster.

Four Would-Be Radical Enhancers: The Technologist, the Therapist, the Philosopher, and the Sociologist

This book's investigation of radical enhancement follows the views of four of its leading advocates. They are Ray Kurzweil, whom I shall label "the technologist," Aubrey de Grey, "the therapist," Nick Bostrom, "the philosopher," and James Hughes, "the sociologist."[7]

Ray Kurzweil, the technologist, is a pioneer in artificial intelligence, and the inventor of the speech recognition technology that enables blind people to use computers. He's also a futurist with a fine record in predicting how technologies will develop and how they will affect us. Kurzweil isn't like the clairvoyant whose apparent success in forecasting the future depends on our generous tendency to forget the many mistaken predictions and remember the few accurate ones. The technological advances that will enable radical enhancement are consequences of a law governing technological change—something Kurzweil calls the law of accelerating returns. According to this law, technologies improve at an ever accelerating rate.

Kurzweil uses three letters to summarize the law's significance for us—GNR. "G" is for genetics. Geneticists have mapped the human genome and begun to connect strands of DNA with human characteristics. They've identified genes that influence how prone to disease we are, our rate of aging, and how intelligent we are. It's the combination of this knowledge with emerging techniques for selecting and altering DNA that Kurzweil envisages making us smarter and healthier.

"N" is for nanotechnology, which involves the manipulation of matter at the atomic or molecular level. Nanotechnology will become a technology of human transformation chiefly by way of microscopic robots called nanobots. Once introduced into human bodies, nanobots will enable enhancements that cannot be achieved by the insertion, deletion, or transposition of the chemical letters of DNA. Some nanobots will purge our arteries of bad cholesterol, while others will fix the glitches in our memories. Still others will create rich and vivid virtual realities that blur the distinction between the real and the virtually real. Kurzweil thinks we will become like the characters in the *Matrix* movies, moving things around in the world by the power of our minds alone. The difference between the movies and the forecast is that these feats won't be achieved in a convincing virtual reality: They will occur in an actual reality that has had the morphing qualities of virtual reality added to it.

"R" is for robotics, the technology that will complete our escape from human biology. Fortunately for us, the G and N technologies will have dramatically enhanced our intellects by the time we are expected to contemplate the robotic revolution, because its implications are beyond current comprehension. With R's arrival, technological change will be so rapid that new technologies will succeed older ones almost instantaneously. We will arrive at *the Singularity*—"a future period during which the pace of technological change will be so rapid, its impact so deep, that human life will be irreversibly transformed."[8] The Singularity isn't just something that might happen. It is the almost inevitable consequence of the law of accelerating returns. And it's going to happen soon. Kurzweil offers 2045 as the year of the Singularity, a prediction that, if true, would make the next forty years considerably more eventful for us and our biological lineage than the previous forty million. He proposes that we will celebrate the Singularity by creating a mind that is "about one billion times more powerful than all human intelligence today."[9]

Kurzweil says two things about this massively intelligent mind. First, it will be human. Second, it will be nonbiological. Its neurons will have been replaced by electronic circuits that are both computationally more efficient and entirely immune from disease. This won't happen all at once—for most of us there's unlikely to be a single replacement event. Kurzweil predicts a gradual merger of human with machine. In its early stages this merger will be motivated by a desire to fix parts of our brains that have become diseased. Cochlear implants already help profoundly deaf people to hear by directly stimulating their auditory nerves. Soon prosthetic hippocampuses could be restoring the memories of people with Alzheimer's disease. Once we install the implants, we will face a choice about how to program them. We hope that they can at least match the performance of the parts of the brain they replace; we hope, for example, that prosthetic hippocampuses will be as good at making and retrieving memories as healthy biological human hippocampuses. But if you've gone to all the trouble of installing a prosthetic hippocampus, why would you rest content with a human level of performance when you could have so much more? From a technological perspective there's nothing sacred or special about our present intellectual powers. Leaving the performance dial set on "human" is a bit like resolving to drive your new Porsche exactly as you drove your old Morris Minor. This attitude to the machinery of thought will lead, in the end, to a complete transformation of the human mind. Chapters 3 and 4 explore Kurzweil's ideas, focusing especially on this idea of uploading ourselves into machines. I will argue that the prudent person will never freely trade in all of his or her neurons for electronic circuits. This should significantly slow our progress toward the Singularity, and perhaps even cancel it.

Predictions about the arrival of ideal states have an annoying habit of not turning out. Speaking in the mid-1950s, the Soviet leader Nikita Khrushchev claimed that Communism, the perfect social arrangement in which there would be no state and people would entirely lack class consciousness, would arrive by 1980. It didn't. Prudent revolutionaries have contingency plans. If Kurzweil's margin of error is anything like Khrushchev's, he may need some help to ensure that he is still around for the Singularity. This brings us to Aubrey de Grey, whom I label the therapist. Calling de Grey a therapist makes him seem like a doctor. But de Grey's aims differ markedly from those of modern medicine. Present-day medical therapies have the purpose of keeping us alive and moderately content for what we think of as a normal human life span. De Grey's therapies may keep us youthful and healthy for thousands of years.

De Grey trained initially as a computer scientist. He became aware that very little was being done about a fact of human existence that is horrible and currently universal—aging. You can be mindful of recklessly driven buses, refrain from smoking, and refuse to holiday in war zones. But these measures only ever postpone death. Even the most careful and fortunate among us face a fate of progressive enfeeblement followed by death. De Grey decided to do something about this and set about teaching himself everything known about the aging process with an eye to reversing it. His goal is to engineer what he calls *negligible senescence*. A negligibly senescent being does not age. As we will see in chapters 5 and 6, his thinking has advanced considerably beyond idle philosophical speculation about how nice agelessness might be. De Grey has a plan. It's possible that he can play a better chess game against death than did the knight in Ingmar Bergman's iconic movie, *The Seventh Seal*.

The details of Kurzweil's and de Grey's proposals are quite complex. However, in both cases there is a central principle that both pulls these details together and aids in their presentation. Kurzweil's master idea is the law of accelerating returns. This law explains why he's confident that we will soon be able to engineer super-intelligence into machines and into ourselves. De Grey's central principle is an idea he calls *longevity escape velocity*. It explains his confidence that some people alive today may still be alive in a thousand years' time. According to de Grey, we need to begin the process of working out how to repair age-related damage as soon as possible. The techniques we invent will add years onto the life expectancies of everyone who has access to them. Longevity escape velocity will have arrived when new therapies consistently give us more years than the time it takes to research them.

Nick Bostrom, the philosopher, does not face technical or biological obstacles. Instead, he confronts arguments that accepting the offer of radical enhancement would be irrational, immoral, or both. Radical enhancement's philosophical opponents include Leon Kass, a conservative social commentator, who has gained fame (and notoriety) for arguing that the right thing to say to the electronic implants and genetic enhancements is "yuck"; Francis Fukuyama, a political economist and historian of ideas, who finds our humanity too fragile to withstand the multiple assaults of enhancement technologies; and Bill McKibben, an environmentalist writer, who opposes the technologies that radical enhancers would direct at human nature with the same energy as he opposes the technologies that currently destroy Mother Nature.[10] These opponents of radical enhancement are known as *bioconservatives*. The word "conservatism" is sometimes used to describe the idea that we preserve certain modes of dress or social arrangements. The brand of conservatism set up in opposition to radical enhancement looks past the superficialities of what we wear and how we relate to one another to our biological fundamentals. Bioconservatives want to spare our humanity from the genetic modifications and electronic implants that they think will destroy it.

The advocates of radical enhancement have an extra reason to worry about the arguments of the bioconservatives. The United States, the country currently best resourced to develop the GNR technologies, is also the place where the bioconservatives have the greatest political influence. Though it would be an exaggeration to say that they have the power of veto over the technologies of radical enhancement, they are certainly well placed to delay their advance.

Chapter 7 explores two responses Bostrom makes to enhancement's philosophical foes. Bostrom's is no timid defense of radical enhancement. He takes the philosophical offensive. In an influential paper written with Toby Ord he accuses many of those who oppose the enhancement of our intellects and the extension of our lives of faulty thinking.[11] According to

Bostrom and Ord, the bioconservative error-in-chief is status quo bias, the idea that one state of affairs is better than another simply because it's the way things are now. In other domains, status quo bias prevents us from making changes that would clearly improve our situation. And, they claim, so it does here. Bostrom and Ord offer a technique that will free our thinking of this error and help us to see the appeal of intellectual enhancement and life extension.

Bostrom's second argument directly addresses the human values that the bioconservatives claim to be defending. He argues that if we really understood our human values we would see that all of us desire the things enabled by radical enhancement. The real debate is between advocates of radical enhancement and those who have failed to understand that their "human" values demand millennial life spans and massive intellects. Bostrom offers a principle intended to bring Kass, Fukuyama, and McKibben out of their bioconservative closets.

James Hughes, the sociologist, addresses the social transformations that the GNR technologies will bring. The dominant way of thinking about the social consequences of enhancement is vividly represented in the 1997 Andrew Niccol movie, *Gattaca*, starring Ethan Hawke and Uma Thurman. It depicts a society in the near future comprised of "valids," those who come from embryos selected for the presence of genes linked with desirable characteristics, and "in-valids," individuals who came into existence in the way that almost all of us do today. The result is a biotech dystopia in which a genetic elite rules with little regard to the welfare of their genetic inferiors.

If you think that there is anything in the *Gattaca* prediction, then you should be extremely worried about the kinds of societies that radical enhancement will create. The differences imagined in the movie are nothing compared with those envisaged by would-be radical enhancers. The valids may be on average slightly healthier and more intelligent than the in-valids, but they are still recognizably human. Hawke's character is an in-valid who outperforms his supposed genetic superiors at a variety of intellectual and physical tasks. The technologies of radical enhancement have the power to create beings whose intellectual and physical gifts are much greater than those of the valids in *Gattaca*. There is next to no chance that an unenhanced human intellect will out-think a machine mind designed to process information over a million times faster than it. If the valids cannot view the in-valids as fellow citizens, then how will the radically enhanced beings that Kurzweil and de Grey want to create regard beings like us?

Hughes argues for a *democratic transhumanism* that addresses these concerns. The citizens of societies organized according to the principles of democratic transhumanism will understand that the vast differentials in power between the unenhanced and the radically enhanced have no bearing on their moral worth. They will view a citizen's IQ or life expectancy as irrelevant to his or her political standing. Hughes thinks that democratic transhumanism can ensure a harmonious future for societies that comprise individuals who are making the transition from humanity to posthumanity at varying speeds, or not at all. In chapter 8 I'll deflate Hughes's optimism.

A Precautionary Approach to Radical Enhancement

This book presents a variety of possible futures that are somewhat darker than those favored by the advocates of radical enhancement. I conjecture that the most dramatic means of enhancing our cognitive powers could in fact kill us; that the radical extension of our life spans could eliminate experiences of great value from our lives; and that a situation in which some humans are radically enhanced while others are not could lead to a tyranny of posthumans over humans.

What should we make of these darker forecasts? They're certainly logical possibilities—they describe futures that radical enhancement could create. But the same is true of the more optimistic outcomes favored by radical enhancement's advocates. How do we separate stories that demand serious attention from those that can be safely dismissed as science fiction?

This book counsels a precautionary approach to radical enhancement. For an indication of what this approach involves, consider the current debate about climate change. Environmental scientists argue that unless we significantly reduce our production of greenhouse gasses, polar ice will melt, pushing up sea levels to such an extent that entire island nations will be submerged, and land that currently feeds a large percentage of the world's population will be rendered barren. In the meantime, we face city-wrecking hurricanes and an accelerating rate of species extinction.

These are not inevitable consequences of the escalating production of greenhouse gasses. The received models of the global climate could be wrong. Even if they aren't, it's possible that we'll find a technological fix for the problem. For example, we may invent nanobots that, when released into the atmosphere, restore greenhouse gasses to their pre–Industrial Revolution levels. In this more optimistic scenario, we avoid global warming and all of its bad consequences without distracting big business from its purpose of making money for shareholders and creating jobs. All that will be required is the commitment of resources to the relevant research in nanotechnology.

Although the various optimistic scenarios about climate change are not impossible, it would be criminally complacent to just suppose that one among them will come to pass. There's a big difference between a plan that has a good chance of being implemented and an idle hope that things won't be as bad as people are saying. Before we abolish carbon credits and other measures designed to reduce the production of greenhouse gasses we should expect to see a pretty advanced prototype of the carbon-eating nanobot. We'd also need proof that it could not only remove carbon but be successfully switched off when target levels had been achieved. We'd need strong evidence that it would not cause additional, more serious problems. The onus is on their opponents to demonstrate that they can make the optimistic scenario—or some relative of it—a reality. Without such proofs we should follow the costly path of curbing our production of greenhouse gasses to give ourselves the best chance of avoiding the even more costly outcome predicted by environmental scientists.

This book's precautionary approach places the onus on radical enhancement's defenders to show that the dark scenarios I describe can be avoided. What can radical enhancers do to ensure that their favored optimistic scenarios are not only possible, but also very probable?

Species-Relativism about Valuable Experiences

Some of the harmfulness of radical enhancement depends on a view about the value of human experiences that I'll call *species-relativism*. According to species-relativism, certain experiences and ways of existing properly valued by members of one species may lack value for the members of another species. In chapter 2, I make the case that radical enhancement is likely to create beings who do not belong to the human species. As we encounter the views of Kurzweil, de Grey, Bostrom, and Hughes, I'll be offering species-relativist arguments for finding purportedly enhanced posthuman existences inferior to unenhanced human existences. Species-relativism, therefore, justifies rejecting radical enhancement.

Species-relativism resembles the cultural relativist view about morality, according to which the truth or falsehood of moral judgments is relative to a culture. Cultural relativists say that the claim that slavery is evil may be true relative to the standards of twenty-first-century New Zealand, but false relative to the standards of the ancient Greek city-state Sparta. According to cultural relativism, it can be right for a member of one culture to refuse to trade her current circumstances and values for circumstances and values presented as objectively superior.

Cultural relativism has few supporters among philosophers.[12] Some of the view's opponents dispute the significance of cultural differences over morality. According to these opponents, a bit of investigative effort usually connects superficial moral disagreements to more fundamental agreements. They make the point that the circumstances of Spartans were very different from those of contemporary New Zealanders. Perhaps if New Zealanders were to find themselves precariously wedged between many perennially hostile neighbors, then they too might resort to and endorse slavery. Correlatively, if the Spartans were to find that they were under no threat of invasion and that sporting rivalries with Australia were their only outlet for collective aggression, then they too might abhor slavery. In each case we change circumstances without changing fundamental moral values.

The fact that the boundaries between different species are more significant than those between different human cultures makes species-relativism a more plausible view than cultural relativism. Some philosophers deny the importance of cultural differences by arguing that our universal human biology is a greater influence on moral views than are the variable cultural circumstances in which we are raised.[13] The importance of species boundaries is not an issue that divides those who take different sides in the long-running nature–nurture debate. Those who oppose cultural relativism because they think that morality is shaped substantially by biology should be open to the idea that the different biologies of different species can generate moral diversity that is both genuine and fundamental.

Species-relativism is not the same as the moral view known as specie-sism justly criticized by Peter Singer and other advocates of nonhuman animals.[14] Speciesists allege that the boundaries between species are morally significant. According to them, humans can often justify treating chimpanzees badly by making the point that chimps are not human. Singer responds that many nonhuman animals are capable of experiencing pain and pleasure and that it's inconsistent to say that these experiences matter a great deal if they happen to humans, but not at all if they happen to nonhumans. Species-relativism does not imply speciesism. Suppose that species-relativism were true of the experience of pain. That could mean that members of our species view pain as something that should be prevented, while members of some rational alien species might not—perhaps the members of this species lack the negative experiences that we call pain and therefore have no concept of them. Although species-relativists may defend the aliens' indifference to *both* chimpanzee and human pain, they can agree with Singer that it's inconsistent for us to say both that it's wrong to cause unnecessary pain to humans and that it's acceptable to do so to chimpanzees.

Does species-relativism offer a plausible account of the value we place on our lives and experiences? Some human values are likely to withstand, and even to be promoted by, radical enhancement. Longer lives and improved intellectual and physical prowess are certainly the objects of human desires; they aren't constructs of transhumanist ideology. The values that correspond with these human desires will doubtless survive our radical enhancement even if we exit the human species. My concern is for the violence done to other human values by the unchecked pursuit of extended lives and enlarged intellects.

For a science fiction example of how radical enhancement could preserve the biological bases of some human values while undermining the bases of others, consider the Cybermen, recidivist invaders of Earth in the BBC TV series *Doctor Who*. The Cybermen began existence as human beings. They followed the path counseled by Kurzweil and extended their life spans and boosted their intellectual and physical powers by replacing flesh with cybernetic implants. By the time the Doctor encounters them, all that remains of the humans they once were are biological brains completely encased by metal exoskeletons.

The Cybermen are a very different end-result of the fusion of human and machine from that predicted by Kurzweil. It would be wrong to overstate the philosophical lessons that can be learned from this example; we shouldn't expect a nuanced analysis of the human condition and possible threats to it from a prime-time TV show. But the Cybermen can nevertheless serve as an illustration of how enhancement might promote certain human values at the expense of others. The self-modifications of the Cybermen are born out of the undeniably human desires to live longer and be smarter. They've achieved these ends, but only by suppressing other aspects of their humanity. For example, the Cybermen require inhibitors to prevent their human emotions from interfering with the directives of logic.

The Cybermen are creatures of fiction, and I won't be supposing that radical enhancement has effects anywhere as extreme as cyber-conversion, the process that turns humans into Cybermen. This book limits itself to the claim that radical enhancement is a way of exiting the human species that threatens many (but not all) of our valuable experiences. Experiences typical of the ways in which humans live and love are the particular focus of my species-relativism. I present these valuable experiences as consequences of the psychological commonalities that make humanity a single biological species. I argue that they are under threat from the manner and degree of enhancement advocated by Kurzweil, de Grey, Bostrom, and Hughes.

The next chapter addresses the tension between radical enhancement and our humanity. Many bioconservatives say that our humanity is the price we have to pay for radical enhancement. Some advocates of radical enhancement agree, effectively wishing our humanity goodbye and good riddance. But others think that we may undergo truly dramatic transformations without losing our humanity. Ray Kurzweil and Aubrey de Grey do not use the term "posthuman," because they think it gives the false impression of a gap between us as we are now and what technology will turn us into. I will argue that radical enhancement is indeed likely to take our humanity from us. The question we must then ask is what is lost along with our humanity.

2   Radical Enhancement and Posthumanity

Radical enhancement involves improving significant capacities to a degree that greatly exceeds what is currently possible for humans. The chief debating examples in the philosophical literature on enhancement involve cases of what I will call *moderate enhancement*. Suppose parents were able to genetically alter their children, making them as smart as the genius physicist Albert Einstein, or as good at tennis as the Swiss maestro Roger Federer. Such cases are correctly viewed as enhancements because they produce capacities considerably beyond the norm for humans—there's a big gap between Einstein and Federer and ordinary physicists and tennis players. But they are moderate rather than radical because they do not exceed the maximum attainable by any current or past human being.[1]

In this chapter, I address an implication of radical enhancement that is obscured in some of the more breezy presentations of the idea. Radically enhanced beings are not only significantly better than us in various ways, they are different from us—so different, in fact, that they do not deserve to be called human. Moderate enhancement raises serious moral issues. For example, it's right to be concerned about the weight of expectation on a boy whose genome has been modified, at considerable expense, with the genetic sequences relevant to Federer's tennis talents.[2] But his parents should be in no doubt about their son's humanity. Federer is a human being, and so would be a human child whose genome was altered to be like his. If the candidates for moderate enhancement enter human, then they should emerge with their humanity intact.

The advocates of radical enhancement are divided on the question of whether our humanity can withstand radical enhancement. Some hold that radical enhancement is compatible with our humanity. These compatibilists include Huxley and Kurzweil. Huxley defends the idea of "man remaining man, but transcending himself, by realizing new possibilities of and for his human nature."[3] One of Kurzweil's central themes is that we remain recognizably human even after we've completely uploaded ourselves into super-powerful computers. He thinks that the beings we will become will share our aesthetic sensibilities and form relationships with one another much like the ones we currently form, including sexual relationships. Nick Bostrom and James Hughes are, by contrast, incompatibilists who emphasize just how different from us radically enhanced beings will be.[4]

A reliable indicator of which camp a contemporary advocate of radical enhancement falls into is the frequency of their use of the word "posthuman," which translates literally as "a being that comes after humans." It appears nowhere in Huxley's writings—though he should probably not be counted because he was writing before it was in popular usage. Kurzweil, on the other hand, makes a point of avoiding the term because he thinks it places a boundary between us and the beings that the GNR technologies will turn us into—a boundary that doesn't exist in reality. He rejects the implication that radical enhancement takes us "beyond humanity." The future that Kurzweil imagines is definitely postbiological, but not posthuman.[5] In the incompatibilist camp, Bostrom presents posthumans as very different from us, comparing the degree of change that we must undergo to become them with that separating humans from australopithecines, tree-climbing prehumans with brains slightly larger than those of chimpanzees.[6]

In this chapter, I argue that though it's not logically necessary that a human who has been radically enhanced will become a nonhuman, it is likely that he or she will. We can then ask what we should make of the loss of our humanity. Incompatibilist advocates of radical enhancement are, unsurprisingly, all in favor. They make the point that it's precisely the respects in which posthumans are better than us that makes them different. If we must renounce our humanity to gain complete immunity from cancer and the ability to know every recorded historical fact simply by downloading it into our enhanced minds, then so be it. The chapter concludes by highlighting significant costs associated with exiting the human species. As we investigate the views of Kurzweil, de Grey, Bostrom, and Hughes in later chapters, we will see just how substantial these costs could be.

Humans as a Biological Species

The obvious starting point for an investigation of whether radical enhancement is compatible with our humanity would be a definition of the concept "human." The bad news is that there is no consensus on what it means to be human.[7] Throughout history humanity has been attributed, or denied, to serve ideological or political purposes. Humans have a predictable and depressing tendency to call "human" those we want to treat well and to deny the humanity of those we want to kill or enslave. It would be quite impossible to reconcile the many different things that people have meant when they invoke humanity.

Since posthumans are my principal concern, I'll pointedly avoid these many controversies about what it means to be human. In what follows I present an analysis of what it means to be human, and consequently of what it might mean to lose one's humanity, that comes from the biological sciences.

I define humans as members of the biological species *Homo sapiens*. A biological species is a group of populations whose members are capable of interbreeding successfully and are reproductively isolated from other groups.[8] On this understanding, humanity is the biological species *Homo sapiens*, comprising individuals capable of breeding with each other while lacking the capacity to breed with members of other biological groups.

How to define a species is a topic of ongoing debate among theoretical biologists and philosophers of biology, and I don't mean to imply that the biological species concept is entirely free of controversy.[9] It is, however, the concept of species most used in contemporary biology. As this chapter progresses, we'll encounter and respond to some problems for this way of defining humanity. I certainly don't pretend that this biological analysis encompasses everything that people mean when they say that *A* is human or deny that *B* is. What I do propose is that it will help us to identify valuable experiences that radical enhancement places under threat.

Once we invoke the biological species concept, the task of saying what a human is becomes more straightforward. But the simplification of our task comes at a cost. Aristotle defined man as the rational animal. Successor definitions added traits such as morality, language use, and tool use to the essence of human beings. These definitions purport to capture something that makes humans more important than earthly nonhumans. The problem they face is that scientists have failed to justify our exclusive claim on attributes such as rationality, language use, tool making, symbolic imagination, and self-awareness.[10] Though the great apes patently aren't human, they do reason, make tools, use language, are self-aware, and may use basic moral concepts.[11]

The biological species concept identifies humans as one biological species among many—it doesn't even pretend to say why humans might be "better" than members of other species. One implication for those contemplating the loss of our humanity is that if humans are not uniquely rational, handy, and capable of speech, then losing one's humanity does not entail the loss of these attributes. Moreover, it becomes hard to understand the rationality of remaining human if there's something better available, namely, posthumanity. The advocates of radical enhancement claim to want to turn rational, moralizing, tool-using humans into creatures who are more rational, more moral, and better at making and using tools.

This apparent gap in the argument against radical enhancement is sometimes plugged by appeals to human nature. For example, according to Fukuyama, it's our shared human nature that is the primary victim of enhancement technologies;[12] he presents this injury to our human natures as dehumanizing us.[13]

I have little to contribute to the debate about human nature beyond offering an observation about why it might have been so difficult to arrive at a single philosophically satisfactory account of it. If we take the biological species concept as our starting point, then we will view human nature as constituted by the large cluster of traits by which one human recognizes another creature as an appropriate mate either for him or herself, or for a sibling or child. There have been millions of years of coevolution between the physiological, behavioral, and psychological cues indicating suitability as a mate and the perceptual and cognitive capacities enabling recognition of and appropriate responses to these cues. As members of our evolutionary lineage have changed, natural selection has delicately recalibrated the weightings of the multiple criteria for good mates and reliable hunt partners. The traits typical of humans feature prominently among these criteria. We recognize chimpanzees and austra-lopithecines as possessing many of the relevant traits—they use tools, have four limbs and two forward-facing eyes, nurse their young, and so on. But we find that they don't possess these traits in quite the right manner, degree, or combination to count as human. This explains our reluctance either to mate with them or to countenance a pair bond between one of them and a child of ours. On the other hand, we have no difficulty in recognizing as human those among us who have suffered disease or injury and have lost cognitive capacities or physical attributes typical of humans. To use a philosopher's term of art, "human nature" is a *cluster concept*. It comprises a range of conditions, none of which is individually necessary or sufficient. Typical humans are bipedal; but we aren't at all tempted to deny the humanity of someone in a wheelchair. Reasoning is characteristic of humans; Alzheimer's disease may destroy this power, but it doesn't exile its victims from the human species. We acknowledge as human individuals who exceed a certain threshold of the relevant properties. We find chimpanzees fascinating because they are so similar to us. But we aren't tempted to call them human, because we recognize them as falling short of the threshold of humanity—they just don't seem similar enough to us, physiologically, behaviorally, or psychologically, to be counted among us.

The account of human nature that emerges from this is not a particularly philosophically tidy one. It's a very long list of typical human characteristics weighted in terms of their significance to our history of finding mates and enlisting allies. It's clear how radical enhancement interferes with the distinctive human combination of these characteristics—radically enhanced beings will be much smarter and longer lived than typical humans. What's not clear, however, is why this should matter. The beneficiaries of radical enhancement will, after all, acquire capacities manifestly superior to those typical of humans.

Understanding Reproductive Barriers

The first question to address is whether someone who comes into existence as a human and then acquires significant capacities well beyond those of any current human being would be reproductively isolated from the rest of us. I suggest that there's a good chance that radical enhancement will in fact create reproductive barriers, and therefore will result in beings who are not human. An answer to this question enables us to address a further, evaluative question. Why might exiting the human species be a bad thing? I will argue that reproductive barriers set the boundaries for an important collection of values, and that we should recognize radical enhancement as infringing on these values.

It might be argued that advances in genetics have largely dissolved the barriers between purportedly reproductively isolated groups of organisms. Allen Buchanan observes that the discovery of cloning by somatic cell nuclear transfer makes asexual reproduction possible for humans.[14] Other techniques permit scientists to routinely swap genetic material between organisms that could never have sex. The same tricks that introduce fish genes into tomatoes could easily add those same genes to human genomes. Buchanan proposes that phenomena such as these render "the notion of 'species boundaries' itself suspect."[15] We should expect scientists with radically enhanced intellects to be even better at combining the genetic material of very different-seeming organisms, thus further reducing the value of the biological species concept in evaluating what might befall humans.

This argument, however, overstates the problem for the biological species concept. Facts about what might or might not happen in a laboratory are actually not directly relevant to decisions about species boundaries. We assign individual organisms to species not on the basis of what happens in human or posthuman laboratories, but according to what occurs in nature. Humans and chimpanzees cannot interbreed in nature, even if they might do so in a specific laboratory setting. Consider an analogous case. The periodic table groups the constituents of matter into chemical elements. Its classifications continue to be applicable even in an age when we can place uranium in a nuclear reactor and turn it into plutonium. Technologies that grant increasing power over the members of the biological species *Homo sapiens* do not show that it's wrong or pointless to group humans into a biological species—at least, they won't until the transformations that they permit become the norm for us.

There do seem to be some individuals who are both human and reproductively isolated from other humans. A man who is surgically castrated cannot reproduce with other humans. Postmenopausal women are no longer capable of reproducing. It's possible that the only sexual activity available to monks confined to monasteries, to murderers condemned to life sentences in all-male prisons, or to individuals deeply impressed by David Levy's book, *Love and Sex with Robots: The Evolution of Human–Robot Relationships*, is non-reproductive.[16] Their reproductive isolation doesn't render them nonhuman. In each of these cases there are past, potential, or counterfactual reproductive connections with other humans. But for his surgery a castrated man could reproduce with other humans. A postmenopausal woman may have had children in the past. Were monks or inmates to leave or escape confinement, or robot-lovers to redirect their romantic attentions, they could reproduce with other humans. These past, potential, and counterfactual connections help them to satisfy the biological species concept of humanity. Furthermore, there's more to reproductive success than sex. A castrated man is denied the option of passing his DNA on to children, but he may share DNA with a brother. He can promote his own biological fitness by volunteering to babysit his brother's children, thereby slightly increasing the chances that a niece or nephew will grow up and find a mate, or perhaps giving his brother an opportunity to go on a date and thereby have more children. These are strategies for getting his genes into the next generation that don't require sex. Similarly, a postmenopausal woman who has already reproduced can boost her reproductive prospects by helping her children to have children of their own.

In each of the cases in the preceding paragraph, other human beings are the vehicles of an individual's evolutionary prospects. In the following pages I'll explore the possibility that this is no longer the case for humans who undergo radical enhancement. Some of the technological advances discussed in this book may give someone who started out as human a destiny that has nothing or very little to do with the human species.

The idea of someone beginning his or her existence as a member of the human species, then undergoing some significant transformation and thereby exiting the species, seems odd.[17] It's clearly not something that happens in the normal course of events. But it does seem at least logically possible for a being to begin its existence as a member of a particular species and to exit that species. The following example will seem like a somewhat contrived analytic philosopher's debating example, but it does help us to understand what might happen in more realistic cases.

There is some debate among anthropologists as to whether *Homo sapiens* and Neanderthals, members of the species *Homo neanderthalensis*, interbred.[18] The Neanderthals were bulky, large-brained hominids who lived predominantly in Europe and the Middle East between 200,000 and 30,000 years ago. According to the received view they became extinct soon after the colonization of their homelands by members of our species. Suppose that the humans and Neanderthals really are two different biological species, and imagine that a team of posthuman genetic engineers and surgeons time-travels to the Europe of 45,000 years ago where they find Neanderthals living near members of our own species. They take a member of the human species and modify his DNA so that his genome is indistinguishable from that of Neanderthals. Posthuman surgeons then alter the human's physiology and psychology to exactly resemble those of Neanderthals. It's plausible that the original individual could survive this sequence of procedures—he could remain conscious throughout. The best way to describe what has happened seems to be to say that an individual started out human and became a Neanderthal. He's no longer attracted or attractive to other humans. He can no longer produce offspring with humans. He is, in compensation, attracted to Neanderthals and can reproduce with them. Furthermore, these offspring are typical Neanderthals who choose to and are able to mate with other Neanderthals. It follows that the biological species concept would describe this process as transforming the individual from a human into a Neanderthal. His evolutionary destiny lies with members of *Homo neanderthalensis* and not with members of *Homo sapiens*.

Now imagine that instead of time-traveling to the Middle Paleolithic the same team of genetic engineers and surgeons come to our time and modify a member of *Homo sapiens*. They make all of the changes that advocates of radical enhancement are asking for. It's surely possible that the changes could move their subject out of our biological species. The question we must answer is which enhancements have this effect.

It might be argued that the biological species concept is not a tool fit for classifying posthumans. In chapters 3 and 4 we will investigate Kurzweil's proposal that we upload ourselves onto computers.[19] Desktop computers don't belong to any biological species, and nor will the entirely electronic intelligences that Kurzweil thinks we will become. In one of the more kinky moments in his book, *The Singularity Is Near*, Kurzweil imagines these intelligences having sex.[20] But whatever mutually pleasurable activities electronic intelligences engage in, biological reproduction is unlikely to be a significant motivation for them. The fact that posthumans may not belong to their original species does not prevent the biological species concept from playing an exclusionary role, helping us to distinguish members of our species from nonmembers. Some things are reproductively isolated from us because they belong to other biological species. Chimpanzees are examples. But others are isolated from us because they are members of no biological species. For example, i-Pods and toasters fall into this category of nonhuman.

Furthermore, reproductive isolation is not all or nothing—it comes in degrees. To see this, consider the plight of the black stilt (*Himantopus novaezelandiae*), a New Zealand bird threatened with extinction. One of the causes of its predicament is, not all that surprisingly, the destruction of its habitat. But, somewhat paradoxically, it is also imperiled by its own reproductive efforts. The species seems to be literally breeding itself to death. Black stilts are increasingly mating with the more plentiful Australian native pied stilt (*Himantopus himantopus*), something that is happening to the extent that black-pied hybrids now predominate in parts of New Zealand. Does the fact that pied and black stilts will mate with one another give the lie to the notion that there are two species of stilts rather than one? This question is not so easily answered. Black and pied stilts are, to some degree, reproductively separated from each other. Pied stilts choose to mate with other pied stilts and black stilts with black stilts when they can, and they select different reproductive targets only when their preferred category of mate is unavailable. The stilts are certainly not reproductively isolated to the same extent as humans and chimpanzees, capable of mating only in the most exceptional, thoroughly artificial circumstances, but this does not mean that reproductive barriers are completely absent. Perhaps time will reduce any reluctance on the part of black stilts to mate with pied stilts. In this event we should acknowledge one biological species comprising individuals with varying colorations. Alternatively, restoration of the black stilt habitat may make them more plentiful, with the consequence that they become much less likely to resort to mating with pied stilts. If this happens then we may move toward a situation in which there are clearly still two distinct biological species of stilts.

The fact that reproductive isolation is not an all-or-nothing affair may sometimes leave unresolved the question of whether two organisms belong to the same biological species. I argue that significant values depend on our integrity as a biological species. Access to these values will become eroded to the degree to which radical enhancement isolates one from the rest of the human species. Reproductive isolation is not all or nothing, and nor are the values founded on it.

The stilt example illustrates something else important about reproductive barriers. Reproductive isolation may be a partly psychological matter. Humans and Neanderthals were similar in many important respects. So, supposing that Neanderthals were a distinct biological species, how were the reproductive barriers between them and us maintained? Perhaps there were genetic or chromosomal incompatibilities that prevented humans and Neanderthals from mating successfully. If we and the Neanderthals had different numbers of chromosomes, then attempts at reproduction could have led either to no offspring, or to offspring that were themselves sterile. We can only guess at the physiological barriers between *Homo sapiens* and *Homo neanderthalensis*, but it is likely that substantial psychological barriers separated the two species. Perfectly healthy Neanderthals will have been somewhat unattractive to beings whose template of attractiveness evolved to match the physiology and physiognomy of healthy humans. Neanderthals were considerably more heavily built than humans. The most striking aspect of their faces was a large beak-like nose, probably an adaptation designed to warm freezing air before it entered the lungs. These traits are likely to have been as repellent to humans as our weedy physiques and bizarrely flat faces will have been to Neanderthals.

Psychological barriers are more significant for humans than they are for most other organisms. There are some organisms for which reproduction is a one-off event. A sexual act brings into existence offspring that have no further contact with either parent. Female freshwater trout lay eggs, males eject sperm, females cover the fertilized eggs with stones, and that is the extent of parental love and commitment of resources. Humans differ in establishing pair bonds whose purpose is to ensure care for offspring. Offspring must be acknowledged as children to qualify for proper care. A Neanderthal–human hybrid is more likely to be perceived as a biological oddity than a bonny wee baby. Its chances of receiving enough love and attention to reach an age when it could have children of its own would have been, as a consequence, significantly reduced. Its chances of attracting a human mate would have also been slim.

We should not overinterpret psychological aversions. The history of our own species contains a depressingly large number of examples of members of one group of humans refusing to mate with the members of another group because they worshiped the wrong gods or had the wrong color skin. These barriers are both insufficiently robust and too temporary to turn humanity into many species. People who lack their group's dominant prejudices can mate successfully with members of the other group. Furthermore, cultural changes such as the spread of new religious or moral beliefs can remove the barrier entirely. Any barrier fit to make a new species must be less permeable and more enduring than those that racism or religion occasionally create.

There is evidence for psychological aversions that are hardwired into humans rather than just by temporary products of culture. Karl McDorman, a researcher from Indiana University, has described an aspect of human psychology that might have evolved to ensure that we don't try to reproduce with things that appear human but aren't really.[21] What has come to be called the *Gollum effect* takes its name from the creepy computer-generated character in the *Lord of the Rings* movies, based on the books by J. R. R. Tolkien and directed by Peter Jackson. Audiences find Gollum repellent. What seems to repel them is his mixture of human and animal characteristics. Gollum's voice is human but his movements are distinctly ape-like. We're repelled by Gollum but not at all by apes or by other humans—our aversion seems to be to the mixture itself. The *New Scientist* article reporting McDorman's findings suggests that the aversion might have evolved to help us to avoid humans with infectious diseases. The reasoning was that the best way to not acquire serious infectious diseases is to not hang out with those infected with them. The Gollum response picks up on the fact that individuals with serious diseases often behave in ways that fall outside of the norms for our species. This seems improbable to me. There are many infectious diseases that have noticeable effects on human behavior but don't produce anything like the Gollum effect. For example, we may have reservations about going in for a snog with someone with bad 'flu, but we don't find them particularly creepy. We don't respond to them in the way *Lord of the Rings* audiences responded to Gollum. It's possible that there's a different reason for the effect. Perhaps it helps us to avoid the suboptimal reproductive consequences of attempting to mate with our evolutionary cousins. This would make the disgust that Gollum evokes in modern movie audiences a descendant of the emotional reaction of humans to Neanderthals.

With the biological species concept in hand, we can ask what kinds of changes might send us into biological exile. I think we can pretty promptly exonerate some purportedly posthumanizing influences. For example,

Fukuyama has argued that drugs like Prozac might inadvertently make us posthuman—or at least they may be a significant step in that direction.[22] He thinks they can do this by eliminating unhappiness, which Fukuyama presents as an essential and valuable part of the human condition. Fukuyama's is no blanket opposition to antidepressants, however. He finds Prozac compatible with our humanity when restricted to its proper therapeutic role, which is the treatment of clinical depression. It threatens our humanity only when prescribed to people currently experiencing normal human levels of happiness and sadness in an attempt to make them even happier.

Fukuyama may have identified a disturbing tendency in the prescription of antidepressant medications. But it is just wrong to think of the person who uses the Internet to construct a list of symptoms of clinical depression to present to doctors in order to get Prozac as petitioning to exit the human species. Prozac does not create reproductive barriers. An already happy person who decides to experiment with Prozac is not thereby prevented from reproducing with members of the human species.

Positioning humanity as a biological species offers a reassuring response to some recent concerns about in vitro fertilization (IVF). After thirty years of IVF babies there is a growing body of evidence that, considered collectively, IVF children differ in certain ways from conventionally conceived children. A recent article offered these differences as evidence that IVF babies, dubbed "IVF-lings," might be a new species.[23] The cover of the magazine in which the story appeared was adorned with the kind of spooky-baby image normally reserved for stories about reproductive cloning. According to the story, when compared with conventionally conceived children, IVF-lings are "taller, slimmer, have higher levels of growth-promoting hormones and their blood lipids (think cholesterol levels) are better."[24] The scientists cited in the article expect to find other differences between IVF-lings and conventionally conceived children. There's actually a big difference between being different and belonging to a new species. Suppose IVF-lings were to be limited to IVF as a means of reproduction. Then we may have the beginnings of a reproductively isolated biological species of IVF-lings. But this is manifestly not the case. For example, Louise Brown, the world's first IVF baby, has a child of her own, conceived the conventional way.

How Radical Enhancement Might Create Reproductive Barriers

We are now in a position to ask whether and how radical enhancement might create reproductive barriers.

The Princeton geneticist Lee Silver was among the first to speculate explicitly that radical enhancement might cause its recipients to exit the human species. He envisages genetic enhancement separating humans into the unenhanced "Naturals" and the "GenRich" beneficiaries of enhancement technologies. In Silver's future history, the GenRich, in their turn, separate into various subgroups delineated along various enhancement agendas—there are GenRich scientists, GenRich businessmen, GenRich musicians, GenRich footballers, and so on.[25] The differences between the GenRich and the Naturals reach to biological fundamentals. Silver predicts that intermarriage between Naturals and GenRich will be infrequent, and he thinks infertility will be a common problem for those rare GenRich–Natural couples.[26]

One reason the members of different species fail to reproduce is that they are genetically incompatible. Gregory Stock offers one explanation for this genetic incompatibility: He speculates that the genes required to radically enhance us might be loaded onto an artificial chromosome.[27] This way of introducing them is supposed to reduce the likelihood of such additional genetic material interfering with existing genes that are essential to survival. The additional chromosome itself could act as a reproductive barrier. The difference in the numbers of chromosomes of horses and donkeys doesn't prevent them from mating. But the mules that result from these unions are almost always infertile. Perhaps adding chromosomes to human genomes might have similar consequences for reproduction. It's possible that the various prostheses that Kurzweil envisages us grafting onto our bodies may create physiological barriers by interfering with the reproductive act. The Cybermen in the TV series *Doctor Who* began as humans but seem, by way of their enhancements, to have exited our species. They boost their numbers not by biological reproduction but instead by subjecting humans to a process known as cyber-conversion, which forcibly and painfully replaces flesh with cybernetic upgrades.

I will not place great emphasis on these chromosomal or physiological barriers. They'll represent nothing more than an engineering problem for specialists in reproductive medicine whose intellects have been radically enhanced. It's pretty likely that electronic implants and genetic enhancements could be placed and administered so that they did not create physiological barriers. Chromosomal differences are, similarly, unlikely to constitute an insuperable barrier. I want to focus on possible psychological barriers that radical enhancement might bring into existence.

In the Transhumanist FAQ, Bostrom castigates those who think that "simply by changing our self-conception, we have become or could become posthuman." Bostrom thinks that "[t]his is a confusion or corruption of the original meaning of the term. The changes required to make us posthuman are too profound to be achievable by merely altering some aspect of psychological theory or the way we think about ourselves. Radical technological modifications to our brains and bodies are needed."[28] Bostrom is surely right that we cannot just redefine ourselves as posthuman. But I think he underestimates the power of self-conceptions. How one conceives of oneself and of others has a significant bearing on whether you would mate with them, and therefore on whether you are reproductively isolated from them.

Suppose you were to travel forward in time to the age of the posthumans. You post a dating advertisement on the Internet (or whatever has replaced it). You give an honest account of your characteristics under the handle of merelyhuman. Your advertisement must compete against that of a posthuman romantic competitor, singularityman, who is determined to highlight the salient differences between your two profiles. I adapt the following profile from what Bostrom says about what it will be like to be posthuman.

I am capable of aesthetic and contemplative pleasures whose blissfulness vastly exceeds anything merelyhuman has experienced. I've reached a much greater level of personal development and maturity than merelyhuman. This is partly because I've been around for hundreds or thousands of years, all of those lived in full bodily and psychic vigor. I'm so much smarter than merelyhuman, he takes weeks and maybe even months to read books that I read in seconds. I see that merelyhuman mentions an interest in philosophy in his profile. As it happens, I am much more brilliant than the most brilliant philosopher listed in his profile. I routinely create artworks, which he could understand only on the most superficial level. Even viewed this way they strike him as wonderful masterpieces. I will be so much better as a romantic partner given that I am capable of love that is stronger, purer, and more secure than he or any human has yet harbored.[29]

I predict two things. This singles advertisement by singularityman will have limited appeal to any human who happens to come across it. Also, your advertisement will have little appeal to singularityman and his ilk. Reproductive barriers will exist between singularityman and human beings, and between merelyhuman and posthumans. These barriers, like those between black and pied stilts, may not be absolute, but they are nonetheless likely to exist.

There's likely to be an enduring and substantial barrier between people whose intellects have been radically enhanced and people with unenhanced intellects. It's unlikely to be as temporary or as permeable as the barriers racism and religion occasionally create. I suspect that, should they be positively disposed toward us, beings whose intellects radically exceed our own will be more disposed to view us as pets than as potential mates.

Other radical enhancements may have more gradual and subtle effects. In chapters 5 and 6 we will explore de Grey's proposal that we should radically extend our life spans. De Grey envisages us living for a thousand years and beyond. Someone whose life expectancy has been radically extended but who is not intellectually enhanced is not prevented from having meaningful conversations with people who have normal human life spans. Any tendencies toward reproductive isolation are likely to be more subtle. Still, people with indefinite life spans are less likely to want to engage romantically with people likely to grow old and die on them.

People living lives of normal human durations might be similarly unimpressed by those with life spans of indefinite duration. For example, de Grey concedes that radical life extension has the potential to lead to overpopulation. A being with an indefinite life span could presumably have indefinitely many children. Those unfazed by the potential multiplication of childcare responsibilities might relish the prospect of being more fecund than Genghis Khan. But de Grey objects to the dramatic increase in the global human population that this way of thinking may lead to. He recommends that we choose between children and indefinite life spans. He has signaled his own moral seriousness by opting for childlessness. It's possible that negligible senescence will reduce the desire for children. One of the motivations for having children is a kind of vicarious survival.[30] Though we may, as individuals, be doomed, we know that some part of us will live on in our children. Negligibly senescent people won't feel such a strong need to vicariously survive in their offspring. Instead, they can look forward to actual, indefinite survival. It's possible this lack of desire for children will make negligibly senescent people less attractive to people whose life spans are not radically extended. I'll have more to say about the psychological consequences of negligible senescence in chapter 6, where I'll argue that people with radically extended life spans will soon become very intellectually and emotionally different from those who continue to age. The reproductive barrier between the senescing and the negligibly senescent is far from absolute, and is unlikely to come into existence immediately. Nevertheless, its existence depends on more enduring and substantial differences than differences in religious and moral beliefs.

According to the view I am outlining, it is not logically necessary that a human being who undergoes radical enhancement will exit our species and either become a member of a different biological species or become a member of no species at all. But it is likely that he or she will. Radical enhancement is likely to create reproductive barriers.

You might ask how different someone who has undergone radical enhancement really is from those who have been rendered infertile, women who have undergone menopause, or have experienced any of the other things that prevent individuals who are undeniably human from reproducing with other humans. These things seem to create reproductive barriers without causing an individual to exit the human species. There are certainly no hard and fast lines, but radical enhancement is properly viewed as significantly different from menopause and castration. Radical enhancement is more likely than either of these events to result in beings who represent some kind of new beginning. Their evolutionary futures are less likely to depend on any contributions that they might make or might have already made to the human species. As we will see in chapter 3, Kurzweil thinks that radical intellectual enhancement directs its recipients toward the Singularity. This possible future is not dependent on what happens to humans who have rejected the path of electronic enhancement. Consider de Grey's plan to make us negligibly senescent. Beings with indefinitely long life spans are likely to be focused on their own continuation rather than on children they might have with humans. Those with definite life spans must reproduce to get their genetic material into the next generation. If they've already reproduced and are incapable of further reproduction, their evolutionary prospects depend on contributions to already existing children, or perhaps to nephews and nieces. Negligibly senescent beings have hit upon a method of ensuring the survival of their genes that does not require reproduction. They will continue to carry their own genes into the indefinite future.

In conclusion, reproductive isolation may not be a logically necessary consequence of radical enhancement. But it is a likely one. Perhaps radically enhanced beings will belong to no biological species. Perhaps they'll belong to many biological species. At the very least, they'll have taken a significant step away from our species. It's now time to address the details of radical enhancement.

3 The Technologist—Ray Kurzweil and the Law of Accelerating Returns

Ray Kurzweil predicts that we'll soon be enhancing our intellects a billion-fold and living as long as we want. These developments are consequences of the Singularity—"a future period during which the pace of technological change will be so rapid, its impact so deep, that human life will be irreversibly transformed."[1] According to Kurzweil, there's little that's optional about the Singularity. It's not something that just might happen. Rather, it's an almost inevitable consequence of the law of accelerating returns, a law that dictates that technologies advance at an ever increasing rate. Kurzweil presents the Singularity as an event in our future that is dragging us toward itself much as a black hole sucks in matter and energy.[2]

The law of accelerating returns most directly affects human beings through its impact on the technologies of artificial intelligence (AI). We tend to view work in AI as principally about designing computers that can think, where progress is gauged by comparing the performances of computers with those of humans. According to Kurzweil, this understates AI's potential significance. Work in artificial intelligence is not just about making artificial things intelligent; it's also about making us artificially super-intelligent. Kurzweil envisages us enhancing our mental powers by implanting powerful neuroprostheses. We'll come to understand that anything that computationally cumbersome, disease-prone neurons and synapses can do, electronic circuits can do better. This will lead to a progressive pensioning off of our biological brains, a process whose completion will see the transfer of minds from brains into machines. We will *upload* ourselves.

This isn't just idle philosophizing. The really impressive thing about Kurzweil is that he's no mere observer of AI. He's the inventor of machines that perform a variety of human thought processes, including understanding speech and reading written language. Kurzweil combines several lucrative computing patents with a knack for predicting developments in AI. On his list of successful forecasts are a variety of military applications of computers, the dramatic expansion of the Internet, and the victories of computers over the world's best human chess players.[3] Bill Gates, who ought to know a thing or two about computers, rates Kurzweil "the best person [Gates knows] at predicting the future of artificial intelligence."[4]

This chapter describes Kurzweil's prediction tool, the law of accelerating returns. The law applies to all technologies, but our chief interest will be the role it could play in furnishing the knowledge necessary to upload humans into machines. We'll examine some obstacles on the way to uploading presented by science writer John Horgan. The chapter concludes with an assessment of Kurzweil's assertion that uploading ourselves is compatible with our humanity, that we can become nonbiological human beings.

The Law of Accelerating Returns

Just about everyone has firsthand experience of the tendency of technologies to become more powerful over time. Users of computers know that last year's cutting-edge machine is this year's children's hand-me-down and next year's doorstop. However, the law of accelerating returns is considerably more than the trite observation that technologies tend, over time, to become more powerful. It's the claim that they become more powerful at an accelerating rate, that technological advance is exponential, not linear.

For an example of exponential increase, consider the sequence of rewards on the television game show *Who Wants to Be a Millionaire?* Contestants in the British version of the show aiming to fulfill the want of its title must give correct answers to fifteen increasingly difficult general knowledge questions. Initially the sequence of rewards follows a linear pattern, increasing in increments of £100. The reward for getting the first question right is £100, for question two it's £200, and for the third question it's £300. If the rewards were to continue to increase at this rate it would take ten thousand questions to reach the eponymous million. But contestants who reach the £300 mark find that, from then on, the sequence of rewards departs from this linear pattern. The next prize is £500, whereupon the increase in rewards adopts an exponential doubling pattern. It moves through £1,000, £2,000, £4,000, £8,000, £16,000, £32,000, £64,000. There is a minor departure from the doubling pattern to achieve the rounder figure of £125,000. The next two questions carry rewards of £250,000 and £500,000. Finally there is the £1,000,000 question, whereupon you might be expected to know that Henry II was the English king married to Eleanor of Aquitaine, or that "googol" is the name of the number represented by the digit one followed by one hundred zeros.[5] The game stops at fifteen questions, which is convenient for Celador, the TV production company that stages the show in Britain and owns the international rights. If the exponential pattern were to continue for ten more questions they would have to rename the show *Who Wants to Be a Billionaire?*

The best-known example of exponential technological change involves Moore's law, named for Gordon Moore, the former chair of the computing giant, Intel.[6] It concerns integrated circuits, indispensable parts of modern computers, in whose development Moore himself played an important role. Integrated circuits are essentially miniaturized electronic circuits made up of transistors. The speed with which they process information, the number of calculations they can perform per second, is determined by how many transistors they contain. In a version of his law proposed in the mid-1970s, Moore observed that the number of transistors squeezed into an integrated circuit doubled every twenty-four months with corresponding consequences for computing power. This pattern of ongoing doubling is exponential rather than linear.

The law of accelerating returns generalizes Moore's insight.[7] Kurzweil proposes that what is true of integrated circuits is true of the information-processing technologies that preceded them. And it's also supposed to be true of allied technologies. Kurzweil cites the accelerating improvement of communications technologies, genome sequencing, magnetic data storage, nanotechnology, and Internet bandwidth. Such is the power of the law of accelerating returns that it encompasses phenomena outside of what we recognize as technology. Kurzweil argues that the evolutionary process fits the pattern described by the law. In evolution, useful characteristics arise by random genetic mutation, are retained by natural selection, and are then built upon by later randomly acquired useful characteristics.

It's important to get clear about what Kurzweil is claiming we can—and cannot—use the law of accelerating returns to predict. The designers of integrated circuits can draw a graph that describes the number of transistors that have been fitted into integrated circuits, recognize it as exponential, and make a pretty good guess about the number of transistors in integrated circuits two years hence. But they don't know exactly how this advance will be achieved. If they did, they would presumably make the improved integrated circuits now, rather than waiting two years to do so.

Kurzweil thinks he knows why others have overlooked this trend. The deceptive aspect of exponential change is that it takes a while to really get going. Consider the doubling pattern exemplified by *Who Wants to Be a Millionaire?* that Moore proposes governs the addition of transistors to integrated circuits. If one starts with a small enough number, the doubling pattern does not seem so different from a linear pattern. Compare the exponential doubling sequence: 1, 2, 4, 8 . . . with the linear "add 2" sequence: 2, 4, 6, 8 . . . . At the outset there is little separating the sequences. It's only after a few iterations that differences begin to become apparent. By the time the linear sequence has hit 20, the exponential sequence has arrived at 512. Five more iterations take the linear sequence to 30 and the exponential sequence to 131,072. And so on.

According to Kurzweil, the law of accelerating returns is behind some recent AI successes. Garry Kasparov is considered by many the best chess player ever. That claim may now require the rider best "human" player—for in 1997 Kasparov was bested by the computer Deep Blue.[8] What made this especially galling was the fact that, ten years prior, Kasparov had boasted "No computer can ever beat me."[9] Perhaps Kasparov was guilty of mistaking exponential improvement for linear improvement. Not only could no computer of 1987 have beaten him, but if the rate of improvement from 1987 to 1997 had only matched that of 1977 to 1987 then it is very unlikely that any computer in 1997 could have beaten him. In fact, the rate of improvement in chess-playing computers between 1987 and 1997 was dramatically greater than that of the previous decade.

Chess-playing is a pretty narrow sample of human intelligence. The important thing for Kurzweil is that Deep Blue's victory is no isolated achievement. We remain supreme at a variety of reasoning and language-processing tasks, but the machines are gaining on us.

How to Build a Human Machine

Kurzweil breaks the task of building a computer capable of thinking like a human into two parts. Part one is to match the human brain's computational power. Kurzweil estimates that the brain's billions of interconnected neurons can perform 10^{16} calculations per second (cps).[10] He boosts this estimate to 10^{19} cps to be sure of capturing all of the detail about connections between neurons that will be required to successfully upload human intelligence into a computer. Kurzweil reckons the human memory at 10^{13} bits of information, where a bit is the smallest unit of information that can be stored on a computer, represented by a single binary number (0 or 1).[11] Again, his interest in eventually uploading human minds into computers leads Kurzweil to boost this number. He proposes that 10^{18} bits should suffice to represent all of the interneural connections required to upload human memories in their entirety. The good news is that we're not far off from matching this power and storage capacity. The law of accelerating returns leads Kurzweil to predict that supercomputers will match the computational power of the human brain early in the 2010s, with one-thousand-dollar machines, the equivalents of today's desktop computers, following a decade later.[12] Advances seem to be keeping to Kurzweil's schedule. In June 2008, IBM announced the supercomputer Roadrunner, created for the purpose of ensuring "the safety and reliability of the nation's nuclear weapons stockpile . . . [and] for research into astronomy, energy, human genome science and climate change."[13] Roadrunner is capable of over 10^{15} cps. IBM engineers brag that a physics problem that Roadrunner could crack within a week would have taken the world's most powerful computer of 1998 twenty years to solve.

There's more to human intelligence than computational power. It's easy to imagine devices fabulously more computationally powerful than the brain that are neither capable of thinking like human brains, nor indeed of entertaining a single thought. A computer would require immense computational power to be able to run complex calculations involving numbers one trillion to the power of one trillion digits long or to exactly calculate the trajectories of all celestial objects detectable from Earth. But it may accomplish this while being as far from a thinker as is an early twenty-first-century ten-dollar pocket calculator. If we're going to make machines that don't merely think, but think as well as we do, then we'll need to master the software of human thought. This is part two of Kurzweil's task. He plans to reverse-engineer the human brain, an enterprise to be accomplished by scanning technologies providing increasingly accurate pictures of the brain's inner workings. Today's magnetic resonance imaging (MRI) devices represent the early stages of this undertaking. They use magnetic fields to generate quite accurate three-dimensional representations of the brain and other body parts. This said, current MRIs miss much of the brain's detail, and Kurzweil envisages the scanning task being completed by miniature robots, or nanobots, introduced into our brains. He asserts that "Once the nanobot era arrives in the 2020s we will be able to observe all of the relevant features of neural performance with very high resolution from within the brain itself."[14]

With detailed maps of human thought we can begin the process of building an electronic duplicate of a human brain—rendering the brain "in synthetic neural equivalents." Kurzweil predicts that we will do this by 2029.[15] At long last, we'll have a computer capable of reliably passing the Turing test, the goal set for AI in 1950 by computing pioneer Alan Turing. A computer that passes the Turing test will be able to converse across a variety of topics in a way that seems entirely human to a human judge. If you visit the longbets Web site you'll find that Kurzweil stakes more than his intellectual reputation on this prediction. He's placed a US$20,000 wager on it.[16]

The Gradual Uploading of Humans into Machines

Suppose advances in brain science do keep to Kurzweil's timetable. We'll soon know enough to render a person's brain in synthetic neural equivalents. The uploaded mind would be more an upgrade than a copy. The electrochemical signals that brains use to achieve thought travel at one hundred meters per second—which sounds impressive until you hear that electronic signals in computers are sent at three hundred million meters per second. This means that an electronic counterpart of a human biological brain will think "thousands to millions of times faster than our naturally evolved systems."[17] Mechanized minds have other advantages over biological brains. They'll be immune from Alzheimer's and other degenerative neurological diseases that blight the latter stages of human lives. Kurzweil makes the further point that "machines can easily share their knowledge. If you learn French or read *War and Peace*, you can't readily download that learning to me."[18]

We may be persuaded by the argument that neurons and synapses are intrinsically dumber than electronic circuits. But abandoning our traditional media of thought for electronic circuits seems a momentous move. It's important to realize that Kurzweil does not envisage our doing it all in one go. He makes the point that the boundaries between humans and machines, once sharply drawn, are beginning to blur. The process of uploading is already underway. Cochlear implants, for example, are surgically implanted devices that enable profoundly deaf people to hear by directly applying electrical impulses to auditory nerves; and there is talk of an electronic hippocampus that will enable people with Alzheimer's to store and retrieve memories as well as the rest of us.[19]

The current focus for designers of neuroprostheses is on compensating for the effects of damage to the brain. But there is no stop sign in nature that prevents us from creating electronic implants that improve on what evolution equipped us with. Suppose you implant an electronic hippocampus into the brain of a patient with Alzheimer's. You hope that it performs at least as well as a healthy biological hippocampus. And if it did, it would satisfy the hopes of most patients. But from a technological perspective, there is nothing sacrosanct about the level of performance typical of a healthy human brain. Recalibrating the device may result in powers of memory beyond that of any purely biological brain.

According to Kurzweil, we and our immediate descendants will increasingly be presented with electronic implants capable of enhancing performance. He envisages implants that will connect their bearers directly to the World Wide Web.[20] Once this happens, any lapses in one's biological memory can easily be remedied by a Google search. The biological parts of this "human–machine intelligence" will be recognized as its weakest links. But we needn't be stuck with them for long. There is not a single function performed by biological brains that cannot, at least in principle, be better performed by electronic circuits. If the law of accelerating returns applies, then these circuits will arrive sooner than we expect.

Kurzweil has a term for those who, for whatever reason, do not renounce biology at a time when others are enjoying the benefits of more efficient and flexible artificial cognition. He calls them *mostly original substrate humans*, or MOSHs.[21] This mode of nomenclature would presumably make those of us who depend on biology to get all of their mental and physical activities done TOSHs, or totally original substrate humans. Kurzweil compares the MOSH or TOSH bias for biology with the nostalgia some express for vinyl records in the era of plentiful and cheap digital recordings. It may be true, he says, that the digital recordings currently available fail to capture aspects of music well rendered by vinyl recording; but once this vinyl "extra" is identified it can be transferred to some digital format.[22] Improvements in digital recording will leave no room for any vinyl "je ne sais quoi." There is, by analogous reasoning, no aspect of human intelligence that cannot be artificially replicated and then enhanced.

Forsaking neurons and synapses for electronic circuits enables an increasingly miraculous, and from our present biological human standpoint, barely comprehensible, series of transformations. Machine–human intelligences who have forsaken neurons and synapses won't hesitate to trade in electronic circuits for even more powerful media of thought. Kurzweil predicts that by the end of the twenty-first century, "the nonbiological portion of our intelligence will be trillions of trillions of times more powerful than unaided human intelligence."[23] This precipitous increase will be enabled by our learning how to exploit the computational potential of matter and energy.[24] Our minds will cannibalize ever-increasing quantities of the previously inanimate universe, reconfiguring it to enhance our powers of thought. Says Kurzweil, "[u]ltimately, the entire universe will become saturated with our intelligence. This is the destiny of the universe. We will determine our own fate rather than having it determined by the current 'dumb' simple, machinelike forces that rule celestial mechanics."[25]

Anyone who's seen the final minutes of Stanley Kubrick's film version of the Arthur C. Clarke novel *2001: A Space Odyssey* should be prepared for the notion that presentations of our evolutionary future aren't necessarily easy for us to grasp. The journey inside the Black Box leading to the "Starbaby" seems to some to be just another example of 1960s movie psychedelia. Perhaps Clarke and Kubrick were on to something, however. The Black Box sequence may be true to the way radical evolutionary transformations are bound to seem to our mere human intellects.

Would the saturation of exponentially increasing quantities of matter by our minds make us immortal? Toward the end of this process we could presumably acquire the option of forestalling the big crunch or any other universe-ending event forecast by astrophysicists. Perhaps we'd do this just by desiring it. Kurzweil explains that "Our mortality will be in our own hands. We will be able to live as long as we want (a subtly different statement from saying we will live forever)."[26] Truly immortal beings don't have the option of suicide, something Kurzweil thinks that we will retain. We might find life without limits boring and choose to end it all, thereby returning all the matter in the universe to the control of the "current 'dumb' simple, machinelike forces that rule celestial mechanics."

In the remainder of this chapter, I pose two questions to Kurzweil. The first concerns whether brain science will keep to the schedule directed by the law of accelerating returns. I explore the locations of two possible hurdles on the way to knowing enough about the human brain to upload it. These hurdles may not be entirely unsurpassable, but they could significantly delay uploading, and therefore push back the Singularity. My second question concerns Kurzweil's suggestion that the law of accelerating returns will give rise to nonbiological humans. I suspect that the machines into which we may transform ourselves are no more deserving of the epithet "human" than are the single-cell organisms from which we originally evolved.

Can It Really Be That Simple?

If Kurzweil's bet about rendering the brain "in synthetic neural equivalents" is a good one, then we'll soon have everything we need to upload our minds into computers. So is it? Someone who might back himself to take Kurzweil's cash is science writer John Horgan. Horgan thinks that Kurzweil's forecast of machines that think like us dramatically undersells the human brain's complexity, the important implication being that we're considerably further from knowing enough to upload ourselves into computers than Kurzweil would have us believe.[27] Computers may be becoming more and more powerful, but we're still a long way off from understanding how to program them to be exactly like human brains.

Horgan makes the point that a healthy adult brain is an enormously complex object, containing "about 100 billion nerve cells, or neurons," each of which "can be linked via axons (output wires) and dendrites (input wires) across synapses (gaps between axons and dendrites) to as many as

100000 other neurons." This all adds up to not billions or trillions, but "quadrillions of connections among its neurons." A quadrillion is a very big number indeed—represented by a one followed by fifteen zeros. To give some sense of how big it is, Horgan explains that "a stack of a quadrillion U.S. pennies would go from the sun out past the orbit of Jupiter." And this only begins to quantify the brain's complexity. Horgan says, "synaptic connections constantly form, strengthen, weaken, and dissolve. Old neurons die and—evidence now indicates, overturning decades of dogma— new ones are born." The neurons themselves are extremely variable. They "display an astounding variety of forms and functions." Added to this are many different varieties of neurotransmitters, neural-growth factors, hormones, and other chemicals that "ebb and flow through the brain, modulating cognition in ways both profound and subtle." The area in which least headway has been made, according to Horgan, is the cracking of the neural code. He explains that "just as computers operate according to a machine code, the brain's performance must depend on a 'neural code.'" This code translates patterns of neuronal firing into "perceptions, memories, meanings, sensations, and intentions." We are only at the very beginning of our understanding how neurons collectively represent the belief that soy milk makes a worse cappuccino than low-fat cow's milk because it doesn't froth quite as well, or the desire to read Edward Gibbon's *The Decline and Fall of the Roman Empire* in order to determine whether there are any similarities between ancient Rome's end and America's present predicament.

Horgan's argument reminds me a bit of the scene in the 1986 movie *Crocodile Dundee*, in which Michael J. "Crocodile" Dundee, a weatherbeaten inhabitant of the Australian Outback played by Paul Hogan, travels to New York and is confronted by a mugger with a switch blade. In response to the warning that his assailant has a knife, Dundee chuckles and says "That's not a knife," whereupon he pulls out a large bowie knife and states *"That's* a knife." Kurzweil can make a similar response to Horgan's attempt to mug him with a large number. Horgan presents a quadrillion as a really big number. But is it *really* a big number? Kurzweil places the brain's computation complexity at 10^{19} calculations per second—that's a one followed not by fifteen but by nineteen zeros. The computer that he thinks will fully accommodate the contents of a human memory will contain 10^{18} bits of information. Such numbers seem truly massive, but they are not beyond the reach of the law of accelerating returns. For example, a doubling sequence starting with two will reach a quadrillion after a mere fifty iterations. If the growth in our understanding of the brain really is governed by the law of accelerating returns, then we should expect it to reach a rate of increase that will put quadrillions of connections well within reach.

But there's more to Horgan's critique than some big numbers. To see why, consider what he says about the neural code. The neural code organizes vast patterns of neuron-firings into the "perceptions, memories, meanings, sensations, and intentions" that we intuitively recognize as constituting the human mind. According to Horgan, early twenty-first-century neuroscience tells us next to nothing about this highest level of brain organization. We know that specific areas of the brain are active when we perceive the faces of loved ones or remember traumatic events. But, for example, we just don't know exactly how, at the level of individual neuron-firings, or networks of neurons, the recognition of the fuel-efficiency and environmental-friendliness of hybrid cars could prompt you to decide to purchase one if you can get a good enough trade-in deal on your 2005 Toyota Corolla. And we're—seemingly at least—a long way from knowing this.

Could it be that we'll never crack the neural code because the human brain's just too complex to be fully understood by humans? This is the view of Peter D. Kramer of *Listening to Prozac* fame, who says that "If the mind were simple enough for us to understand, we would be too simple to understand it."[28] Obviously, if the brain is unknowable by us then we'll never know enough to upload it. Kurzweil is not at all impressed by this defeatist skepticism, and the points I'm about to make don't rely on it.[29] In what follows I'll make the case that even if the neural code is knowable, we might still be a long way away from knowing enough to render the brain "in synthetic neural equivalents." I'll explore two possible reasons it may take longer than Kurzweil thinks for us to know enough about the human brain to successfully upload it. These obstacles may cause us to fall significantly behind the schedule dictated by the law of accelerating returns.

A Speed Bump on the Road to Uploading

The first possible delay arises in connection with the relationship between the neural code that organizes vast patterns of neuron-firings into the "perceptions, memories, meanings, sensations, and intentions" and the activities of individual and small groups of neurons that are the principal focuses of contemporary neuroscience. There are two views in the philosophy of science about the relationship between the high-level things like perceptions and memories and low-level things such as neurons and synapses.

Atomism is the idea that complex systems can be exhaustively and completely analyzed in terms of their parts. For example, ecological atomists think that ecosystems can be completely described in terms of things happening to and being done by the totality of their living and nonliving elements. On this view, ecologists could fully describe all of the individual organisms in a given ecosystem, say everything about their physical circumstances, and thereby have nothing else to say about the ecosystem. *Holism* is the denial of this idea, sometimes summarized by the folk wisdom that the whole is greater than the sum of its parts. If ecological holism is true, one could completely describe all of an ecosystem's living and nonliving elements and still have important facts about the ecosystem left to discover.

Kurzweil is certainly not committed to the stereotypical atomistic view of brain science that would hold that a complete inventory of neurons and connections between them will yield all of the brain's secrets. He thinks that our knowledge of the brain should proceed in both directions—top-down and bottom-up. He writes, "Brain reverse-engineering will proceed by iterative refinement of both top-to-bottom and bottom-to-top models and simulations, as we refine each level of description and modeling."[30]

We can use Horgan's discussion of the neural code to pinpoint the location of a speed bump on the journey to a completed neuroscience. Brain scanners don't directly reveal the perceptions, memories, meanings, sensations, and intentions that Horgan suggests constitute the neural code. They're things that neuroscientists will have to infer from activity witnessed on MRIs, from information beamed to them by brain-scanning nanobots, and from their observations of the behavior of the system as a whole. If holism is true, then a completed theory about the things that MRIs and nanobots scan won't inevitably yield a complete account of the neural code, for the simple reason that perceptions, memories, meanings, sensations, and intentions are more than the sum of the activities of neurons and neural networks and maps. Upload technicians equipped with a complete picture of the brain at the level of individual neurons, networks, and neural maps could fail in their attempts to transfer human minds into machines. They may successfully replicate the detail of human brains at the level of neurons, networks, and neural maps but not succeed in relating them in ways that produce perceptions, memories, meanings, sensations, and intentions. This failure would be due to the fact that the proper relations between neurons, networks, and maps are only visible at a level of activity that we haven't yet properly grasped. Suppose the upload technicians were to combine the elements in such a way that they exactly matched the activity of a living brain over a given period of time. They might be pretty confident that the upload is thinking. But, deprived of the neural code, they might have no such confidence about any novel patterns. They would rightly wonder whether the novel patterns correspond with new thoughts or are just violations of the neural code best classified as meaningless gibberish.

We can compare their predicament with that of someone listening to a conversation conducted in a language of which he is completely ignorant. Suppose the listener exactly transcribes the sounds of five minutes of the conversation. If he managed to utter these sounds with perfect accuracy he could be confident that he's successfully spoken the unknown language. But, as he is deprived of knowledge of language's semantics and syntax, his confidence should not carry over to any novel combinations. Sounds that seem to him to be relevantly similar to those produced by the native speakers are likely to be meaningless babble to them. Children regularly conduct this experiment when they imitate the sounds of speakers of foreign languages. English-speaking children's versions of "Chinese" or "German" may fool their parents, but they never fool native speakers. Those trying to build thinking machines face analogous difficulties if they proceed without knowledge of the neural code.

Neurological holism doesn't lead ineluctably to Kramer's skepticism—that we're guaranteed never to be smart enough to know everything there is to be known about our brains. To see why not, think again of the parallel with ecological holism. Ecological holists deny that we can find out all there is to know about an ecosystem by studying its parts. But many of them do think that we can find out about the higher-level organizational principles of ecosystems. They insist that formulating these will require a kind of holistic perspective. Ecologists will need to mentally step back so as to see the forest rather than just all of the individual trees. If this mental maneuver is possible for human ecologists, then there's no obvious reason neuroscientists couldn't adopt the same holistic perspective and so formulate the high-level principles that collectively completely characterize the neural code. The speed bump arises in respect of how we'll arrive at any new fundamental principles governing the neural code.

If neurological atomism is true, then we can be reasonably confident that our understanding of the brain will be completed at an ever increasing rate, aided by exponential improvements of scanning and other technologies. This is not so if neurological holism is the case. If holists are right about the neural code then the gaps in our knowledge won't be filled by increases in the power of the technologies we use to investigate the brain. Technological improvements should lead *part of our investigation* of the brain to advance at an ever-increasing rate. But the most miraculous scanners will be of little value if the people who are operating them don't know how to locate higher-level structures in the details provided by scans. They may need some new fundamental principles to tell them how to translate various constellations of neuron-firings into perceptions, memories, meanings, sensations, and intentions. Though these new principles may not be beyond discovery by humans, there's no reason to believe that the pace of their discovery is governed by the law of accelerating returns, that it necessarily proceeds at an exponential rate. In the past, the discovery of fundamental principles has relied on the intuition and creativity of brain scientists. Early twenty-first-century neuroscientists are not eight or sixteen or thirty-two times more creative and insightful than Santiago Ramón y Cajal and Camillo Golgi, the geniuses who founded neuroscience in the 1890s. Though they do know a great deal more about neurons and synapses than did Ramón y Cajal and Golgi, there's no reason to think they should be much better at formulating new fundamental principles of brain organization. Perhaps the breakthrough that will lead to a complete description of the neural code awaits not the next iteration of the improvement of brain-scanning technologies, but instead a new Ramón y Cajal, or a new Golgi.

Once we've succeeded in reverse-engineering human intelligence, the law of accelerating returns may enable us to predict the arrival of successive quantum leaps in genius. But before that time there's unlikely to be a strict timetable for the arrival of human minds up to the task of formulating new fundamental ideas. It follows that we have no way of predicting the discovery of the principles governing the neural code.

Kurzweil doesn't think that we need any new fundamental principles to tell us about mysterious high-level neural codes. He reassures us that "The human brain is a complex hierarchy of complex systems, but it does not represent a level of complexity beyond what we are already capable of handling."[31] We can be confident of this because we are able to run accurate simulations of parts of our brains. Kurzweil explains that "it is only recently that [mathematical models of our brains] have become sufficiently comprehensive and detailed to allow simulations based on them to perform like actual brain experiments."[32] Holists won't be reassured by this. To their ears, Kurzweil's assurances are like those of atomistic ecologists who give a long list of ecosystem elements that they've completely understood and then announce "mission accomplished." The atomists may say that they can design computer simulations that correctly predict the activities of all of the elements of an ecosystem. But holists will allege that the combination of these simulations can easily fail to capture the proper relations of the elements. By analogous reasoning, a seemingly accurate mathematical simulation of parts of the brain may leave out key aspects of the neural code governing the brain as a whole. There are likely to be many ways to relate well-functioning neurons and properly behaved neural networks that don't correspond with actual thinking.

Why the Road to Uploading May Be Longer Than We Think

We've just explored a possible location for a speed bump on the way to knowledge sufficient to successfully upload ourselves. I now consider a further reason for delay. Kurzweil may have significantly underestimated the length of the road that leads to uploading. The view that I'll describe is compatible with neurological atomism. It's motivated by a different view about what the basic elements of a completed theory of the brain should be.

Judgments about how close we are to knowing enough to upload the brain depend on how much there is to know about it. Note that we don't have to make this kind of assessment when evaluating progress in integrated circuits. We measure increases in computational power based on the numbers of calculations a given integrated circuit can perform per second. Unlike with the study of the brain, there's no computational end point that we think of ourselves as aiming at. We're just trying to make chips better and better . . . and better.

The point I'm making is a simple one. Recall the doubling sequence in *Who Wants to Be a Millionaire?* Now consider a different TV quiz show— *Who Wants to Be a Quadrillionaire?*—that follows the same doubling sequence and whose host takes the same amount of time to pose questions. Contestants in the imaginary show potentially spend longer in the hot seat than do contestants in the actual show simply because a greater number of doublings separate £1,000 from £1,000,000,000,000,000 than separate £1,000 from £1,000,000.

In this example, the names of the respective quiz shows specify their end points. The problem in neuroscience is that no one tells us in advance how many questions about the human brain we have to be able to answer correctly to know enough to reverse-engineer it, and therefore to be able to upload ourselves. Horgan insists that there's a huge amount that remains to be discovered. He gestures toward levels of organization of which we are currently barely cognizant. Though it may seem that we've learned a great deal since Ramón y Cajal and Golgi in the late nineteenth century, suppose that Horgan's conjectures are correct and that neuroscience as of 2010 captures a mere 2 percent of the brain's workings. We should expect the exponential doubling of our knowledge to yield a complete picture in slightly fewer than six hundred years. Kurzweil will be pinning his hopes on the research of Aubrey de Grey, the life extension scientist whose ideas we'll explore in chapters 5 and 6.

You might ask how it could be that Kurzweil has egregiously underestimated how much we need to know about the brain to upload it. Considered in its totality the brain is actually a highly complex object. A fully exhaustive account must take account of many different levels of activity. At the top level are the perceptions, memories, meanings, sensations, and intentions that Horgan talks about. At lower levels of brain organization are neural maps and circuits comprising collections of neurons. There's a level of organization populated by individual neurons and the synapses and dendrites that effect connections between them. Beneath that there's a molecular level of organization comprising the constituents of neurons. Further down still is the subatomic level populated by bosons, mesons, and other entities that continue to perplex physicists. And so we continue, all the way down to matter at its most fundamental level, if indeed there is one. The task of providing an exhaustive account of the brain at all levels of organization could potentially be without end.

Kurzweil's optimism about the imminent arrival of a model of the brain sufficiently detailed to permit uploading depends on the recognition that much of the detail about how the brain works can be eliminated from our model of it. There's actually a good reason that you don't hear neuroscientists talking much about bosons and mesons. They've resolved that the activities or these and other subatomic particles don't need to be included in a model of the brain. This resolution seems entirely principled. Consider how someone might go about designing a computer model of a city's public transport system. A highly accurate model is likely to eliminate many details about the city's buses, trams, trains, and taxis. Facts about the subatomic properties of individual buses and bus drivers will not be included, for the simple reason that knowing about them doesn't really contribute to our understanding of the transport system. Kurzweil is confident that much of the brain's detail can be omitted from a model of it. We'll be able to eliminate subatomic information. And, he claims, we'll also be able to eliminate some information about the behavior of individual neurons. Kurzweil says of the models of neurons and synapses that "The models are complex but remain simpler than the mathematical descriptions of a single cell or even a single synapse. . . . these region-specific models also require significantly less computation than is theoretically implied by the computational capacity of all the synapses and cells."[33]

Eliminating low-level detail will vastly simplify the task of modeling the brain. However, although it's eminently sensible to eliminate detail, my question concerns how Kurzweil can know, in advance of a completed theory about the neural code, *which details* about the brain are dispensable. What is the source of Kurzweil's confidence that his models of the brain are pitched at the correct level of detail, that they include absolutely everything we'd want to upload onto a machine? Consider the ongoing mystery about precisely how the brain generates consciousness. While there's certainly no shortage of hunches about the brain and consciousness, few would pretend to have answered all the questions about how the activities of neurons generate the rich and vivid experience of remembering, say, one's first day at school. It's possible that the current difficulties are due, in part, to the fact that neuroscience's current lowest level just doesn't go deep enough.

The eminent physicist Roger Penrose has advanced a dissident view about how the brain generates consciousness.[34] According to this view, consciousness results from quantum computation effected by structures within neurons called microtubules. It goes without saying that Kurzweil rejects this view.[35] But at this point in proceedings, his rejection can only really come down to a hunch. We don't yet have a complete theory of consciousness, so we can't be sure that quantum computation isn't required to explain it. The resulting view could be atomistic in that it explains all the high-level phenomena in terms of lower-level goings-on; it just so happens that these lower-level goings-on extend to the quantum level. Penrose's theory of consciousness is compatible with Kurzweil's point about the potential eliminability of some of the brain's detail from a model of it. It leads to a different view about *which detail* can be eliminated. The inclusion of quantum physics in brain science is bound to make the task more difficult. Neuroscientists are more like contestants in *Who Wants to Be a Quadrillionaire?* than in the more easily completed *Who Wants to Be a Millionaire?*

I'm not staking anything on Penrose's views about microtubules and quantum computation. We'll have a better idea about who's right in this exchange when conventional neuroscience has advanced to the point when it can yield detailed theories of conscious experience. If there's been no significant improvement on the current somewhat sketchy theories, then we may have to take Penrose's view, or some other account pitched at the subatomic level, seriously.[36] My point is that, right now, we're adjudicating a clash of hunches about the future of neuroscience. There's stuff about the brain that remains to be explained. Perhaps we can explain it all in the way Kurzweil anticipates; perhaps we can't. If the second alternative turns out to be true, we'll need to learn a good deal more about the brain before we go uploading ourselves onto computers.

Kurzweil's Slippery Slope Argument for Human Machines

Suppose we do learn enough about the human brain to upload it. Is this something that we should want to do? I turn to this question in the remainder of this chapter and in the following one.

Kurzweil's vision of the future seems, initially at least, to be the most alien and alienating of all the predictions about a radically enhanced future. We will cease to be biological beings and become super-intelligent machines. You might wonder what the difference is between a version of these events that has us becoming super-intelligent and the version according to which humans become extinct and are replaced by super-intelligent robots. Kurzweil is firmly of the opinion that the future he describes is ours. The robots haven't replaced us; rather, we have become them. Kurzweil thinks the artificial intelligences that will have emerged from us will be "human even if they are not biological."[37] He recommends the label "human–machine intelligence" to emphasize this fact.

What kind of claim do completely nonbiological beings have on humanity? Purely electronic minds seem much less like us than other creatures we have no difficulty in distinguishing from humans. For example, in many ways we seem more similar to chimpanzees and Neanderthals than we do to the robotic beings described by Kurzweil.

Kurzweil's chief argument for the humanity of machine intelligences points to the gradual nature of their emergence from beings who are uncontroversially human. We have already begun to augment the functioning of human brains with electronic components;[38] cochlear implants help profoundly deaf people to hear. There will soon be other implants that restore function to diseased brain and body parts. Kurzweil poses a series of rhetorical questions to bring home the philosophical significance of this slippery slope from biological to nonbiological humans:

If we regard a human modified with technology as no longer human, where would we draw the defining line? Is a human with a bionic heart still human? How about someone with a neurological implant? How about someone with ten nanobots in his brain? How about 500 million nanobots? Should we establish a boundary at 650 million nanobots: under that, you're still human, and over that, you're posthuman?[39]

Kurzweil thinks that an answer in the negative is so obvious as to not be worth stating. We're human at the beginning, and we'll be human at the end. To deny this claim would require us to identify the particular nanobot or neurological implant that takes our humanity from us. No individual nanobot or implant seems sufficiently important to be pronounced the modification that turns a human into a posthuman.

Slippery slope arguments like this one should arouse suspicion. Consider a familiar, somewhat hackneyed philosophical example involving baldness. It is undeniable that some people are bald. It is equally undeniable that some people are not bald. One goes bald by a process that involves losing individual hair follicle after individual hair follicle. There is no single hair follicle that makes the difference between baldness and hirsuteness. But this does not mean that there is no difference between being bald and being hirsute. The best thing to say is that there is a region of vagueness in which one is somewhere between being clearly bald and being clearly hirsute.

By analogous reasoning, just because there may be intermediate cases that are difficult to classify does not mean that the distinction between machine and human is not conceptually clear.

The 1972 movie *Beware! The Blob* featured an alien organism that has arrived on Earth. The blob absorbs the body of its prey, which is transformed into the gelatinous substance that constitutes the creature. Once this meal is assimilated the blob becomes correspondingly larger and rolls on to absorb its next target. Consider what might happen to you if you were attacked by the blob. At the beginning of the digestive process you're unmistakably human. At its termination what was once human is just another part of the blob. In between the beginning and the end of the digestive process are stages that are difficult to describe. They could be a human being nearing the end of its existence, or a not quite fully processed part of the blob, or perhaps both of these things at once. This slippery slope argument should not lead to the conclusion that blobs are humans or that humans are blobs. Even if there are intermediate cases that are difficult to describe, there is a clear difference between the concepts "human" and "blob." By similar reasoning, even if there may be intermediate stages in the process Kurzweil imagines us undergoing that are neither straightforwardly human nor straightforwardly machine, this does not show that there is no difference between an intelligent human and an intelligent machine.

Kurzweil tries other ways to convince us of the humanity of machine intelligences. He proposes that machine intelligences will deserve the label "human" because their design will be based on us, even if the stuff they're made out of is entirely different. Kurzweil's claims about human design are true for the first neurological implants. They are designed to be integrated into human brains and are reverse-engineered from the parts of the brain they are meant to replace. For example, cochlear implants can't work unless they successfully mimic the function of sensory receptors known as hair cells that stimulate the cochlea. If the earliest fully electronic minds are essentially cobbled together neurological implants whose design is based on biological minds, then it seems likely that they will think like us. According to Kurzweil, there may be a deliberate decision to retain our emotional concerns and aesthetic sensibilities as the design of electronic minds evolves. He says:

Even with our mostly nonbiological brains we're likely to keep the aesthetics and emotional import of human bodies, given the influence this aesthetic has on the human brain. (Even when extended, the nonbiological portion of our intelligence will still have been derived from human intelligence.) That is, human body version 3.0 is likely to look human by today's standards, but given the greatly expanded plasticity that our bodies will have, ideas of what constitutes beauty will have expanded over time.[40]

I think this is very unlikely. What may be true for the first electronic minds is unlikely to be the case for later models. We reverse-engineer cochlear implants from human brains because they must enter into a functional relationship with neurons and synapses. For similar reasons, the first electronic hippocampuses will be slavishly copied from well-functioning biological hippocampuses. But there's a good chance that the design of machine minds will soon diverge from the evolved principles of biological design. I suspect that our aesthetic and emotional sensibilities will be among the first aspects of our biological legacy to be ditched. This is because there's a connection between our aesthetic sensibilities and emotional responses and our biology. For example, our ideas about human beauty are shaped by a tacit recognition about what kind of person counts as a good mate. We feel reassured by the sound of rain on a tin roof because we enjoy the experience of feeling protected from it. Completely nonbiological beings will have no need to use physical beauty as an indicator of genetic health, simply because their mode of reproduction, if indeed they do reproduce, will not involve passing on genes. Their countermeasures to bad weather will be altogether more high-tech than tin roofs. Retaining human aesthetic sensibilities may seem as eccentrically anachronistic as a human's resolving to be guided in the selection of a partner by the kinds of characteristics that would have appealed to australopithecines. Our ancestors ditched these aesthetic sensibilities in response to changes in the appearances of mates. Posthuman machine intelligences are likely to have different needs, and their emotional and aesthetic sensibilities will be attuned to these.

There's another option for Kurzweil. Perhaps it doesn't matter that machine intelligences are not human; they may nevertheless preserve the minds of individuals who are currently human. Kurzweil may be wrong to believe that we will retain our humanity through the various transformations. But so long as it's us—human or not—at the end of the total replacement of neurons and synapses by electronic circuits, this process will be correctly characterized as one in which we relinquish our humanity to become something different and better. Perhaps this idea that we can survive as machines is no more conceptually absurd than the religious proposition that we can survive as disembodied souls.

Kurzweil presents a view of our essential properties that would permit us to survive uploading. According to *patternism* we are essentially patterns that can be realized either biologically or electronically.[41] The uploading process's preservation of the relevant patterns ensures that our memories, beliefs, and other mental states are transferred from our biological brains to the electronic medium. We survive.

So, perhaps Kurzweil should give up on the idea that the transformations he hopes for from the GNR technologies will preserve our humanity; this may not be such a big issue so long as they preserve *us*. I address the issue of whether it's rational to upload in the next chapter.

4 Is Uploading Ourselves into Machines a Good Bet?

The law of accelerating returns specifies that technologies become more powerful at an ever-increasing rate. Far from being exceptions to the law, our minds are among its most important applications. The message from AI is that anything done by the brain can be done better by electronic chips. According to Kurzweil, those who grasp this message will progressively trade neurons for neuroprostheses. When the transfer of mind into machine is complete, our minds will be free to follow the trajectory of accelerating improvement currently tracked by wireless Internet routers and portable DVD players. We'll soon become millions and billions of times more intelligent than we currently are.

In this chapter, I challenge Kurzweil's predictions about the destiny of the human mind. I argue that it is unlikely ever to be rational for human beings to completely upload their minds onto computers—a fact that will be understood by those presented with the option of doing so. Although we're likely to find it desirable to replace peripheral parts of our minds—parts dedicated to the processing of visual information, for example—we'll want to stop well before going all the way. A justified fear of uploading will make it irrational to accept offers to replace the parts of our brains responsible for thought processes that we consider essential to our conscious experience, even if the replacements manifestly outperform neurons. This rational biological conservatism will set limits on how intelligent we can become.

For the purposes of the discussion that follows, I will use the term "uploading" to describe two processes. Most straightforwardly, it describes the one-off event when a fully biological being presses a button and instantaneously and completely copies her entire psychology into a computer. But it also describes the decisive event in a series of replacements of neurons by electronic chips. By "decisive" I mean the event that makes electronic circuits rather than the biological brain the primary vehicle for a person's psychology. Once this event has occurred, neurons will be properly viewed as adjuncts of electronic circuits rather than the other way around. Furthermore, if Kurzweil is right about the pace of technological change, they will be rapidly obsolescing adjuncts.

The precise timing of the uploading event is more easily recognized in the first scenario than it is in the second. It's possible that there will be some vagueness about when electronic circuits, rather than the biological brain, become the primary vehicle of a person's psychology. The uploading event may therefore be spread out over a series of modifications rather than confined to a single one.[1] Toward the end of this chapter I will suggest how we might recognize at what point the gradual replacement of neurons with electronic circuits has the effect of either transferring a person's psychology from the former to the latter, or destroying it altogether.

Strong and Weak Artificial Intelligence

One reason Kurzweil is enthusiastic about uploading is that he's a believer in *strong AI*, the view that it may someday be possible to build a computer that is capable of genuine thought. Computers already outperform human thinkers at a variety of tasks. The chess program on my PC easily checkmates me, and my guesstimates of the time are almost always wider of the mark than is the reading on my PC's clock. But the computer accomplishes these feats by means of entirely noncognitive and nonconscious algorithms. The law of accelerating returns enables Kurzweil to predict that work in AI will lead to computers that genuinely think instead of just performing some of the tasks currently done poorly by human thinkers. He believes that refinements in hardware and software will transform today's number crunchers and word processors into the machine minds of 2029.

There's an alternative view about the proper goal of artificial intelligence. Advocates of *weak AI* think that computers may be able to simulate thought, and that these simulations may tell us a great deal about how humans think. They maintain, however, that there is an unbridgeable gap between the genuine thinking done by humans and the simulated thinking performed by computers. Saying that computers can actually think makes the same kind of mistake as saying that a computer programmed to simulate events inside of a volcano may actually erupt. The law of accelerating returns will lead to better and better computer models of thought. But it will never lead to a thinking computer.

Kurzweil has more than an academic interest in this debate. He needs strong AI to be the correct view because what we say about computers in general we will also have to say about the electronic "minds" into which our psychologies are uploaded. If weak AI is the correct view, then the decision to upload will exchange our conscious minds for entirely non-conscious, mindless symbol manipulators. The alternatives are especially stark from the perspective of someone considering uploading. If strong AI is mistaken, then uploading is experientially like death. It turns out the light of conscious experience just as surely as does a gunshot to the head. If strong AI is the correct view, then uploading may be experientially like undergoing surgery under general anesthetic. There may be a disruption to your conscious experience. But then the light of consciousness comes back on and you're ready to proceed toward the Singularity.

In what follows I outline a debate between Kurzweil and an opponent of strong AI, philosopher John Searle.[2] I present them as asking humans who are considering making the decisive break with biology to place a bet. Kurzweil proposes that we bet that our capacities for conscious thought will survive the uploading process. Searle thinks we should bet that they will not. I will argue that even if we have a high degree of confidence that computers can think, you should follow Searle. Only the irrational among us will freely upload.

We can see this bet works by comparing it with the most famous of all philosophical bets—Blaise Pascal's Wager for the prudential rationality of belief in the existence of God. The Wager is designed for those who are not absolutely certain on the matter of God's existence, that is, almost all of us. It leads to the conclusion that it is rational to try as hard as you can to make yourself believe that God exists. Pascal recommends that people who doubt God's existence adopt a variety of religious practices to trick themselves into belief.

Pascal sets up his Wager by proposing that we have two options—belief or disbelief. Our choice should be based on the possible costs and benefits of belief and disbelief. The benefits of belief in God when God turns out to exist are great. You get to spend an eternity in paradise, something denied to those who lack belief. The cost is some time spent in religious observances and having to kowtow to priests, rabbis, imams, or other religious authorities. If you believe in God when there is no God, you miss out on paradise; but then so does everyone else. True disbelief brings the comparatively trifling benefits of not having to defer to false prophets or to waste time in religious observances. The payoffs in matrix form are given in table 4.1.

The Wager is supposed to give those people who currently think that God's existence is exceedingly unlikely a reason to try as hard as they possibly can to make themselves believe in God. The reward for correctly believing is infinite, meaning that it's so great that even the smallest chance of receiving it should direct you to bet that way. Doubters could look upon faith in the same way as those who bet on horses might view a rank outsider that happens to be paying one billion dollars for the win. But God is a better bet than any race horse—there's no amount of money that matches an eternity in paradise. This means that only those who are justifiably certain of God's nonexistence should bet the other way.

My purpose in presenting Pascal's Wager is to establish an analogy between the issues of the prudential rationality of belief in God and the prudential rationality of uploading your mind onto a computer. What I refer to as "Searle's Wager" moves from the possibility that strong AI is false to the prudential irrationality of uploading.

Uploading is an option that is not yet available to anyone. Kurzweil thinks that we'll have computers with human intelligence by 2029 and uploading will presumably be technologically feasible only sometime after that. This means that we're speculating about the decisions of people possibly several decades hence. But our best guesses about whether people of

Table 4.1

	God exists	God does not exist
Bet on God	**You gain a lot**—You receive the infinitely valuable reward of an eternity in paradise.	**You lose a little**—You waste time worshiping god(s) and must defer to religious authorities.
Bet against God	**You lose a lot**—You miss out on the infinite reward of an eternity in paradise. You may suffer the infinite punishment of an eternity in hell.	**You gain a little**—You ignore religion's demands. You have extra time to devote to more pleasurable or meaningful activities.

the future will deem uploading a bet worth making has consequences for decisions we make now about the development of artificial intelligence. If we're confident that uploading will be a bad bet we should encourage AI researchers to direct their energies in certain directions, avoiding more dangerous paths.

Kurzweil versus Searle on Whether Computers Can Think

To understand why we should bet Searle's way rather than Kurzweil's, we need to see why Searle believes that computers will be forever incapable of thought.

Searle's argument against strong AI involves one of the most famous philosophical thought experiments—the Chinese Room. Searle imagines that he is locked in a room. A piece of paper with some "squiggles" drawn on it is passed into him. Searle has no idea what the squiggles might mean, or indeed if they mean anything. But he has a rule book, conveniently written in English, which tells him that certain combinations of squiggles should prompt him to write down specific different squiggles, which are then presented to the people on the outside. This is a very big book indeed—it describes appropriate responses to an extremely wide range of combinations of squiggles that might be passed into the room. Entirely unbeknownst to Searle, the squiggles are Chinese characters and he is providing intelligent answers to questions in Chinese. In fact, the room's pattern of responses is indistinguishable from that of a native speaker of the language. A Chinese person who knew nothing about the inner workings of the room would unhesitatingly credit it with an understanding of her language. But, says Searle, it is clear that neither he nor the room understands any Chinese. All that is happening is the manipulation of symbols that, from his perspective, are entirely without meaning.

What Searle says about the Chinese Room he thinks we should also say about computers. Computers, like the room, manipulate symbols that for them are entirely meaningless. These manipulations are directed, not by rule books written in English, but instead by programs. We shouldn't be fooled by the computer's programming into thinking it has genuine understanding—the computer carries out its tasks without ever having to understand anything, without ever entertaining a single thought. Searle's conclusions apply with equal force to early twenty-first-century laptop computers and to the purportedly super-intelligent computers of the future. Neither is capable of thought.

Defenders of strong AI have mustered a variety of responses to Searle.[3] I will not present these here. Instead I note that Kurzweil himself allows that we cannot be absolutely certain that computers are capable of all aspects of human thought. He allows that the law of accelerating returns may not bring consciousness to computers. According to Kurzweil, the fact that "we cannot resolve issues of consciousness entirely through objective measurement and analysis (science)" leaves a role for philosophy.[4] Saying that there is a role for philosophy in a debate is effectively a way of saying that there is room for reasonable disagreement. This concession leaves Kurzweil vulnerable to Searle's Wager.

One reason we may be unable to arrive at a decisive resolution of the debate between Kurzweil and Searle is that we aren't smart enough. In the final stages of Kurzweil's future history we (or our descendants) will become unimaginably intelligent. It's possible that no philosophical problems will resist resolution by a mind that exploits all of the universe's computing power. But the important thing is that we will be asked to make the decision about uploading well before this stage in our intellectual evolution. Though we may then be significantly smarter than we are today, our intelligence will fall well short of what it could be if uploading delivers all that Kurzweil expects of it. I think there's a good chance that this lesser degree of cognitive enhancement will preserve many of the mysteries about thought and consciousness. There's some inductive support for this. Ancient Greek philosophers were pondering questions about conscious experience over two millennia ago. Twenty-first century philosophers may not be any more intelligent than their Greek counterparts, but they do have access to tools for inspecting the physical bases of thought that are vastly more powerful than those available to Plato. In spite of this, today's philosophers do not find ancient Greek responses to questions about thought and consciousness the mere historical curiosities that modern scientists find ancient Greek physics and biology. Many of the conundrums of consciousness seem connected to its essentially subjective nature. There is something about the way our thoughts and experiences appear to us that seems difficult to reconcile with what science tells us about them. It doesn't matter whether the science in question is Aristotle's or modern neuroscience.

Searle's Wager

Searle's Wager treats the exchanges between Searle and his many critics in much the same way that Pascal's Wager treats the debate over God's existence. The availability of uploading will present us with a choice. We can choose to upload or we can refuse to. There are two possible consequences of uploading. The advocates of strong AI think that the computers we are uploaded into are capable of conscious thought. If Kurzweil is right, you will not only survive, but your powers of thought will be radically enhanced. If the doubters are right, then uploading is nothing more than a novel way to commit suicide.

Suppose we lack certainty on the question of who is right. We must place a bet. We can bet that Kurzweil is right and upload, or we can bet that Searle is and refuse to. For simplicity's sake let's suppose that if you accept the offer to upload yourself into the computationally superior electronic medium then you must consent to the destruction of the now obsolete biological original. It's not hard to imagine candidates for uploading consenting to this. Keeping the original around will seem to machine super-intelligences a bit like leaving an australopithecine former self hanging around to occasionally bump into and be embarrassed by. Later I will explore the implications of uploading while leaving the biological original intact.

I propose that if you are not certain that Kurzweil is correct it is irrational to upload. My reasoning is summarized in table 4.2. You may be about as confident that Kurzweil is right as you can be about the truth of any philosophical view, but if there is room for rational disagreement you should not treat the probability of Searle being correct as zero. Accepting, as Kurzweil does, that there is reasonable disagreement on the issue of whether computers can think leaves open the possibility that they cannot. This is all that the Wager requires.

I should point out that the Wager does not actually require a nonzero chance of Searle's argument being sound. It makes use only of his conclusion. Suppose Searle's argument turned out to be contradictory. The probability of its being sound would be zero, even while the proposition that forms its conclusion "No computer can think" has a positive probability of being true. Contradictory arguments can have true conclusions. The fact that Searle's argument is the focus of vigorous philosophical debate should

Table 4.2

	Kurzweil is right. The Upload is both you and capable of conscious thought.	Searle is right. The Upload is incapable of conscious thought.
Choose not to upload	**(A) You live** You benefit from enhancements that leave your biological brain intact. You miss out on other more significant enhancements.	**(B) You live** You benefit from enhancements that leave your biological brain intact. You are spared death and replacement by a nonconscious Upload.
Choose to upload and destroy your biological brain	**(C) You live** You benefit from enhancements available only to electronic minds. Your life is extended. Your intellect is enhanced. You are free of disease.	**(D) You're dead** You are replaced by a machine incapable of conscious thought.

be viewed as, at the very least, raising the probability of the statement "No computer can think" turning out to be true.

Why Death Could Be So Much Worse for Those Considering Uploading Than It Is for Us Now (Or Why D Is So Much Worse Than B)

Pascal envisages our betting behavior being influenced by the fear of missing out on heaven. In Searle's Wager, the fear of death is operative. Perhaps there's a difference between these penalties that affects their power to motivate a bettor's choice. Death is a very bad thing for most of us, but it's not as significant a loss as missing out on an eternity in paradise. Indeed, for some people it may not be much of a loss at all. For example, a person about to expire from cancer can choose between certain death from disease and a merely possible death by uploading.

This way of thinking about Searle's Wager makes it seem like another bet placed by those who want to live considerably longer than the three score years and ten-plus some that most people currently content themselves with. The bad news for fifty-year-old would-be millenarians is that they are condemned to die according to humanity's traditional schedule unless there's considerable progress in life-extension technologies. Fortunately there's a backup plan, one explicitly endorsed by Kurzweil, and Aubrey de Grey, the theorist to be discussed in chapters 5 and 6. This is cryonics.

Those who sign up for cryonics hand over some money and then authorize the personnel of their chosen provider to take charge of their bodies as soon as they are legally pronounced dead. Their bodies are transported to a cryonics facility whereupon they are cooled to −196 degrees centigrade, the boiling point of liquid nitrogen. They await restoration to life at such time as we know both how to cure the illness listed on their death certificates as having killed them and how to successfully warm a human body from a vitrified −196 degrees centigrade to a living, breathing +37 degrees.

Surprisingly, there's considerable ambivalence about cryopreservation even among those scheduled to undergo it. There's none of the enthusiasm that typically accompanies various techniques of life extension and cognitive enhancement. In fact, some of its defenders reckon cryopreservation "the second worst thing that can happen to you."[5] But they find it clearly preferable to the worst thing—death. There are many eventualities that may prevent clients from being reanimated. For example, the facility may forget to pay its power bill. It may go into receivership and have its plant, equipment, and "stock" bought by a company that supplies practice materials for medical students. The civilization that possesses the means to bring the frozen dead back to life may have no interest in reanimating decadent types from the era of wanton environmental destruction. But the truth is that cryonics may be a good bet even if the probability of reanimation is very small. Consider the following argument presented by Ralph Merkle, a director of the Alcor Life Extension Foundation, a cryonics facility in Arizona:

[T]his small but finite chance is incomparably greater than the chances of revival after a cremation or burial (which are zero). To cryonicists, the chance of resuscitation is worth the money required to fund cryonic suspension arrangements. If resuscitation proves impossible, they reason, you are no "deader" than you would have been without suspension, so what have you lost?[6]

Merkle summarizes the reasoning behind his wager in table 4.3.

Table 4.3

	Cryonics works	Cryonics doesn't work
You sign up for cryonics	You live	You're dead
You don't	You're dead	You're dead

This table does slightly overstate the appeal of cryonics. Cryonics laboratories charge a fair amount of money. In 2008, Alcor charged a minimum of US $150,000 for a "whole body suspension," and $80,000 for a neuro-suspension, in which only the head is preserved. Instead of being given to Alcor, this money could be spent on a Porsche, something with more certain benefits. But Merkle's point is nonetheless well made. If you invest in cryonics you have the potential to benefit hugely for a cash outlay that some people will find pretty trivial. If you instead spend the money on the Porsche, your chances of experiencing the twenty-second century drop significantly.

The potential payoffs for uploading differ significantly from those for cryonics. The bets are similar in that their rewards are contingent on future technological developments. If Kurzweil is to be believed, uploading will become available sometime after 2029. According to one Internet cryonics FAQ, those who undergo cryopreservation today should expect to be reanimated some time between 50 and 100 years from now.[7] But there's a difference in terms of the timing of the placement of the uploading and the cryopreservation bets. Cryonics is something you can do now. Those who are interested should probably contact their preferred provider as soon as possible to minimize the chances of inadvertent burial or cremation. Although you can certainly declare your interest in uploading today, the key choice or choices are likely to come sometime after 2029, when the technologies necessary to transfer the key elements of your psychology onto a machine have been invented.

The difference in the timing of the placement of the cryonics and uploading bets matters. Candidates for cryopreservation face the stark choice between death followed by burial or cremation on the one hand, and vitrification possibly followed by reanimation and treatments for currently incurable illnesses on the other hand. The choice for candidates for uploading is different. They are unlikely to find themselves stricken with terminal cancer and prepared to give the procedure a go. According to Kurzweil's timetable, the candidates for uploading should be experiencing many of the benefits of the genetics revolution, the first in the GNR series of technological revolutions. Aubrey de Grey, the topic of chapters 5 and 6, has explored what might be done for humans without having to resort to electronic upgrades. If he's right about the near future of our species, we could soon become ageless, guaranteed never to suffer from cancer, heart failure, or any of the other diseases that might incline us to disregard caution about uploading. We could be anticipating an indefinite number of additional healthy, youthful years. Genetic enhancements may have considerably boosted our intelligence and aesthetic sensibilities. And genetics is unlikely to be the only source of enhancements. Our intellectual and physical abilities could be boosted by nanobots and neuroprostheses deemed compatible with the survival of our biological brains.

My argument here does not depend on the viability of de Grey's claims. Rather, it relies on the likelihood of their being achieved *relative* to the achievability of uploading. Uploading requires not only a completed neuroscience—total understanding of what is currently the least well-understood part of the human body—but also perfect knowledge of how to convert every relevant aspect of the brain's functioning into electronic computation. As we'll see in chapter 5, the first significant milestone for de Grey is not total victory over human aging, but rather longevity escape velocity. LEV doesn't require full and final fixes for heart disease, Alzheimer's, cancer, and the other conditions that currently shorten human lives. What's essential is just that we make appreciable and consistent progress against them. De Grey's first milestone is technologically more straightforward, and likely, therefore, to come sooner than uploading.

I conclude that people presented with the option of uploading are unlikely to find that they've got little to lose should the procedure fail to transfer their minds into machines. They'll be loath to renounce the variety of enhancements compatible with the survival of their biological brains.

Why Uploading and Surviving Is Not So Much Better, and Possibly Worse Than Refusing to Upload (Or Why A Is Not So Much Better Than and Possibly Worse Than C)

Perhaps you'll miss out on a great deal if you choose to upload and Searle happens to be right about the procedure's consequences. But the potential gains could be truly massive. Uploading opens up enhancements much more dramatic than those made possible by the comparatively few nanobots and neuroprostheses properly deemed compatible with the biological brain's survival. If Kurzweil is right, freed of biological limitations we can become vastly more intelligent. The gap between enhancements compatible with the survival of the brain and those enabled by uploading and incompatible with its survival is likely to be very great indeed.

To begin with, the uploaded mind will be more an upgrade than a copy. The electrochemical signals that brains use to achieve thought travel at one hundred meters per second; this sounds impressive until you hear that electronic signals in computers are sent at three hundred million meters per second. This means that an electronic counterpart of a human biological brain will think "thousands to millions of times faster than our naturally evolved systems."[8] But there's a more theoretical reason why Kurzweil believes that "[o]nce a computer achieves a human level of intelligence, it will necessarily soar past it."[9] Computers are technology, improvements of which are governed by the law of accelerating returns. Although biological brains may improve over time, they're subject to the dramatically slower, intergenerational schedule of biological evolution. Kurzweil thinks that machine minds will learn how to fully exploit the computational potential of matter and energy.[10] They will cannibalize ever-increasing quantities of the previously inanimate universe, reconfiguring it to enhance their powers of thought. Recall Kurzweil: "[u]ltimately, the entire universe will become saturated with our intelligence. This is the destiny of the universe. We will determine our own fate rather than having it determined by the current 'dumb' simple, machinelike forces that rule celestial mechanics."[11]

Suppose we accept that the enhancements compatible with the brain's survival are likely to be significantly more modest than those enabled by uploading. What should those who are considering uploading make of this gap?

Consider the measures taken by economists to convert the objective values of various monetary sums into the subjective benefits experienced by individuals. For most of us, a prize of $100,000,000 is not 100 times better than one of $1,000,000. We would not trade a ticket in a lottery offering a one-in-ten chance of winning $1,000,000 for one that offers a one-in-a-thousand chance of winning $100,000,000, even when informed that both tickets yield an expected return of $100,000—$1,000,000 divided by 10 and $100,000,000 divided by 1,000. The $1,000,000 prize enables you to buy many of the things that you want but that are currently beyond you—a Porsche, a new i-Pod, a modest retirement cottage for your parents, a trip to the pyramids, and so on. Many of us also have desires that only the higher reward will satisfy—a mere million won't buy a luxury Paris apartment or a trip to the International Space Station aboard a Soyuz rocket, to give just two examples. So we have no difficulty in recognizing the bigger prize as better than the smaller one. But we don't prefer it to the extent that it's objectively better—it's not one hundred times better. The conversion of objective monetary values into subjective benefits reveals the one-in-ten chance at $1,000,000 to be significantly better than the one-in-a-thousand chance at $100,000,000.

I think that the subjective significance of the gap between enhancements compatible with the brain's survival and those incompatible with it is unlikely to match its objective magnitude. In fact, there may not be too much of a gap to those considering uploading between the appeal of the objectively lesser enhancements compatible with the survival of our brains on the one hand, and the objectively greater enhancements enabled by uploading on the other. The more modest enhancements will satisfy many of the desires of people for whom uploading is an option. They will live significantly longer, they can be freed of disease, they can play better bridge, learn foreign languages with ease, and so on. Uploading may enable feats well beyond those achieved by people whose biological brains are supplemented with only those electronic chips deemed compatible with the brain's survival, but we have comparatively few desires that correspond specifically with them. There's a reason for this. Desires are a practical mental state—they motivate people to act in certain ways. Those who want to get fit do exercise; those who want to lose weight go on diets; and so on. Desiring radical enhancement is a matter of placing your faith in the law of accelerating returns and waiting for something quite miraculous to be done to you. There's little you can do about it now beyond reading and rereading *The Singularity Is Near* and enrolling in your local chapter of World Transhumanist Association.

One way to adjust the subjective values of $1,000,000 and $100,000,000 is to make the choice from a standpoint of considerable wealth. Donald Trump is likely to look on the smaller sum as barely enough to achieve anything worthwhile. The larger sum, on the other hand, may suffice to acquire some significant piece of Manhattan real estate. This could lead him to prefer the one in a thousand chance to acquire $100,000,000 to the one in ten chance to receive the mere million. Is there an analogous move that can be made in respect of enhancement? It's true that those who've already achieved super-intelligence are likely to be more impressed by the objectively greater enhancements possible after uploading than we are. The problem is that it's impossible for us to adopt this standpoint. We're necessarily deciding about uploading in advance. Compare: We might be able to imagine Trump's contempt for a mere million dollars; but in advance of actually acquiring his wealth we're unlikely to be motivated by this imaginary contempt.

I suspect that the problem may actually be more serious than that signaled by the previous paragraphs. The manner of radical cognitive enhancement permitted by uploading may be worse than the more moderate variety compatible with the survival of our brains in the light of some of our more significant desires. Many of the things that we desire may be contingent on our current level of cognitive powers. We want to protect our relationships with our loved ones. We want to promote and honor our strongest moral and political ideals. Radical enhancement may not remove our capacity to protect, promote, and honor these commitments. But it may remove our desire to do so. Concern about doing the things we currently most want to do may, therefore, lead us to place a low value on radical cognitive enhancement. I will have considerably more to say about these and related human values in chapter 9.

What does all this mean for those presented with the option of uploading? I suspect candidates will prefer more modest, safe enhancements to those whose potential magnitude is greater but come with a risk of death.

Can We Test the Hypothesis That Uploading Preserves the Capacity for Conscious Thought?

Would it be possible to test the hypothesis that someone could survive uploading? If tests do become available and yield the right results, then candidates for uploading may be more confident that they'll survive the process.

Suppose neuroscientists identify the parts of the brain responsible for mathematical calculation. They might carefully excise them from your brain, perfectly preserving them so that they could be reintroduced after the experiment. The removed parts would be replaced by an electronic implant programmed to perform calculations using techniques that are similar to those used by humans. You are then requested to perform some calculations. The experimenters would ask you whether or not you are conscious of multiplying 5 by 5 to arrive at 25. Have you noticed any differences between your current and your former experience of multiplication?

It's not clear that you would notice any such difference even if Searle is right about how this implant works. This is because there's unlikely to be any distinctive variety of conscious thought associated with multiplying. There is, in contrast, a distinctive kind of awareness associated with arriving at the answer 25 and preparing to say "5 times 5 is 25" but this awareness would be shared in the case in which the answer arrives by neuronal computation and the case in which the neuroprosthesis presents it to consciousness. Perhaps there are other opportunities for differences in awareness. For example, you may be aware of stages on the way to arriving at the answer. Suppose you consciously decide to arrive at the answer by the cumbersome route of starting with 5 and adding four lots of 5 to it. You may become conscious of a series of intermediate stages in the calculation. For example, having resolved to add four lots of 5 you become aware that the first addition gives the answer 10, that the second gives 15, and so on. But it would be hard to distinguish the hypothesis that the implant's calculation is a form of thinking from the hypothesis that it is noncognitively number-crunching and presenting bulletins of its workings to the conscious part of your mind.[12]

It's possible that the experience of gradually replacing parts of your brain with electronic prostheses would be a little like neurodegeneration. Alzheimer's disease progressively destroys parts of its sufferers' brains. People tend to become aware of the consequent loss of cognitive skills only when they fail to perform tasks that were once well within their grasp. Suppose Searle is right about the impossibility of machine thought. The gradual replacement of systems in your brain with electronic equivalents would be like Alzheimer's in one significant respect. It would involve the progressive erosion of your powers of thought. The difference would be that you would lack an external check on this loss. You would be like a person with Alzheimer's who has an electronic assistant that informs her of the names of people who she has forgotten. The Alzheimer's patient is probably aware of the assistant. You, in contrast, may be entirely oblivious of the operation of your implanted assistant. The fact that you would perform all the tasks you used to be able to perform would deprive you of any awareness of the ongoing shrinkage of your mind.

Perhaps there's another way to test the hypothesis that uploads can think. We could ask them. There are likely to be plenty of people to ask. Just because it's irrational to upload doesn't mean people won't do it. People do many things that are irrational.

"Uploads" will do a great many things strongly indicative of conscious thought. They will insist that they are conscious. In his response to Searle, Kurzweil presents these claims as a strong pragmatic justification for accepting the consciousness of the computers we will become, or will attempt to become:

these nonbiological entities will be extremely intelligent, so they'll be able to convince other humans (biological, nonbiological, or somewhere in between) that they are conscious. They'll have all the delicate emotional cues that convince us today that humans are conscious. They will be able to make other humans laugh and cry. And they'll get mad if others don't accept their claims.[13]

Purged of its question-begging language, this is exactly what Searle expects. In his thought experiment, something that works like a computer exactly matches the linguistic behavior of native speakers of Chinese without understanding a single word of Chinese. He allows that a computer installed in a robot might perfectly replicate human conscious behavior without ever entertaining a single conscious thought. The machines' skill at convincing people is no more a hallmark of truth than is a subprime lender's capacity to convince home purchasers that they'll cover the mortgage repayments with ease.

The computers that have replaced those who have chosen to upload may be sincere in their protestations of consciousness—either that or they could perfectly simulate sincerity. Human beings who are being insincere are typically aware that they're acting deceptively. There would be no such awareness in your computer replacement. Suppose someone were to challenge your claim that you are conscious. This insult might prompt you to introspect on the workings of your mind, an activity that leads you to vigorously assert that you really are conscious. Now consider the uploaded version of you. It will go through the processes that are the exact machine equivalents of this introspective gathering of evidence. But if Searle is right, its conviction that it is conscious will be the consequence of entirely nonconscious, nonrational manipulations of symbols.

Is it really so unlikely that computational efficiencies mandated by the law of accelerating returns will have detrimental consequences for consciousness? Consider one of the chief benefits of the switch from electrochemical to electronic processing. Kurzweil presents the latter as dramatically faster than the former. An electronic copy of your mind that made no improvements to the ways in which you approached problems would think a million thoughts for your one. It's possible that this gain in speed imperils your conscious experiences. There is much that our brains do that we are unaware of. They compute the angles and velocities of tennis balls enabling us to just reach out and catch them. They make adjustments to the positioning of our bodies that prevent us from falling over. They translate a diverse collection of perceptual inputs into the impression of a baby crying. The thing that these otherwise very different kinds of mental processing have in common is that they are very swift relative to conscious thought processes. Could there be an absolute upper limit in processing speed above which consciousness cannot be sustained? The mystery of consciousness precludes a definitive answer to this question. If the answer is yes, then Kurzweil's proposal to speed all of our thinking up may completely banish consciousness from our minds. None of the kinds of thoughts that we currently entertain will last long enough to be conscious. Of course one might obviate this particular concern by slowing down the electronic circuits. But this would eliminate much of the motivation for the conversion from electrochemical to electronic computation.

Why We Shouldn't Make Electronic Copies of Ourselves

So far I've argued for a relatively restricted claim. It is unlikely to be rational to make an electronic copy of yourself and destroy your original biological brain and body. This leaves open the possibility of making the copy and *not* destroying the original, an option that might seem to offer the best of both worlds. Humans do not need to risk death. They can copy themselves into machines that will then be free to follow the law of accelerating returns all the way to the Singularity.

This raises questions about the relations between what Kurzweil refers to as mostly original substrate humans—those that are substantially biological—and beings that are either super-intelligent, or who act as if they are. In what follows I use the prefix "quasi" to summarize Searle's view about the intelligences that will emerge from the law of accelerating returns. Accordingly, if Searle's right then our future could be shared with quasi-super-intelligent beings. I will draw pessimistic conclusions about these relations in chapter 8. For now I restrict myself to a few warnings about the dangerous implications for human beings of the exponential increases in computing power predicted by Kurzweil.

The following paragraphs are somewhat speculative. They concern the implications for MOSHs of the more distant consequences of the law of accelerating returns. The parts of Kurzweil's books that describe these consequences are the hardest for his intellectually unenhanced readers to understand. They really are the printed page equivalents of the "inside the black box" sequence at the end of Kubrick's *2001: A Space Odyssey*. It isn't surprising that we, unenhanced humans, have difficulty connecting with them, given that they concern beings either far more intelligent than us, or who act as if they are. It's difficult to make predictions about what super-intelligent or quasi-super-intelligent beings will decide to do. But it's important that we attempt to do so. If advances in the GNR technologies were to follow a linear pattern then we should expect to have millennia to evaluate and possibly avoid any unpleasant consequences of machine super-intelligence or quasi-super-intelligence. But if the law of accelerating returns applies, then AI technologies will become more powerful at an exponential rate, and this means that super-intelligent or quasi-super-intelligent artificial intelligences could be with us soon. Our last opportunity to take action to prevent their arrival may be even sooner.

In what follows I propose that we should try to prevent the creation of autonomous or quasi-autonomous electronic minds that are either more intelligent than us or act as if they are. If uploading creates such beings then it should be banned.

Eliezer Yudkowsky confronts the problem of hostile artificial intelligences directly. He thinks we must try hard to create "friendly" AI, which he defines as "the production of human-benefiting, non-human-harming actions in Artificial Intelligence systems that have advanced to the point of making real-world plans in pursuit of goals."[14] Anyone with more than the most casual acquaintance with science fiction is likely to be more familiar with unfriendly artificial intelligences than with the friendly variety. Think of the Skynet of the *Terminator* movies, the Cylons of the

Battlestar Galactica TV series, or the malevolent artificial intelligence of the *Matrix* movies. Data, the friendly android in *Star Trek: The Next Generation,* is decidedly in the minority. Yudkowsky thinks that the key to harmonious relations between humans and super-intelligent artificial intelligences is the attitude of the seed AI. The seed AI is, according to Yudkowsky, the first artificial intelligence capable of self-improvement—to use Kurzweil's terminology, it's the first AI that's capable of self-directedly evolving toward the Singularity. Yudkowsky proposes that we should make sure that the seed AI is friendly. He hopes that successor artificial intelligences will inherit this friendliness. It will act as a constraint on their actions, much in the way that morality acts as a constraint on human actions.

I'm not optimistic about the success of Yudkowsky's project. In chapter 8 I will address issues that arise when trying to predict the moral views of beings who are dramatically more intelligent than us. I restrict myself here to some preliminary comments. I find it hard to imagine that (quasi-) super-intelligent artificial intelligences won't escape the constraints of human-programmed friendliness.

There are two ways in which we might imagine AIs committing unfriendly acts. Such acts could be motivated by goals that are explicitly unfriendly. Examples are the desires to sadistically torture humans or to go on a shooting rampage. Some explicitly immoral desires in humans may result from psychological dysfunctions that programmers could exclude from the seed AI. Of greater concern are implicitly unfriendly goals, goals that are not explicitly unfriendly, but may require unfriendly acts. Consider the desire for money. People get money in ways that don't involve acts unfriendly to other humans. But some ways to get money do involve such acts. Those who convince themselves that procuring money is more important than complying with the rules of morality manage to overcome any squeamishness about theft or murder.

An AI's desire to continue to enhance its intelligence (or quasi-intelligence) is not explicitly unfriendly. But it may nonetheless require unfriendly acts; it may be implicitly unfriendly.

Suppose we were to create machines that are more intelligent than us. Kurzweil predicts that electronic intelligences will evolve to the point at which they will swap electronic circuits for even more powerful media of thought. At this point, intelligence will begin to saturate the universe's matter and energy. By saturation, Kurzweil means that future intelligences will use "the matter and energy patterns for computation to an optimal degree" based on their knowledge of the physics of computation. Intelligence (or quasi-intelligence) will saturate ever increasing areas of the universe. Some parts of the universe will be inhabited by MOSHs, that from the perspective of the artificial intelligences, are making computationally suboptimal use of their constituent matter and energy. I'm not sure what the experience of having the matter and energy that constitute your brain and body absorbed and computationally optimized by a superpowerful artificial intelligence would be like. But it doesn't sound pleasant.

We might try to prevent this by making moral rules more important to AIs than they are to us. Suppose AIs are either autonomous or quasi-autonomous. I suspect that they will find ways around moral restrictions programmed by humans. If you doubt this consider our own case. Natural selection has designed our minds and bodies with the purpose of passing on our genes. To achieve this we have been instilled with a desire for sex. This desire is quite powerful. Our impulse to reproduce can lead us to make life-altering choices without properly evaluating consequences. Unwanted pregnancies are an indicator of the power of nature's programming. But we are rational. We can act against this programming if we judge doing so to be in our best interests. For example, we can use contraceptives. Natural selection is not entirely without countermeasures when faced with this rational avoidance of reproduction. The desire it has implanted in our minds can lead us to have sex too hurriedly—or too drunkenly—to properly apply condoms. But it remains the case that we can get the sex without achieving natural selection's goals. If we are sufficiently rational to subvert natural selection's programming then it seems probable that super-intelligent, autonomous (or quasi-super-intelligent, quasi-autonomous) artificial intelligences will also work out ways to get around our programming of them. They will (quasi-)autonomously make their own (quasi-) decisions about which goals are so important that they require their programmed inhibition against harming humans to be overridden or worked around.

We can be fairly confident that AIs originating from people who freely choose to upload will be convinced of the value of the intellectual enhancement that comes as they approach the Singularity. These artificial intelligences will be strongly motivated to work around any moral restrictions that might impede their progress toward the Singularity. If they're much better at solving problems than we are, then we should expect them to be much better than us at circumventing preprogrammed restrictions on their behavior. They're unlikely to do the AI equivalent of forgetting to use condoms.

In the Pixar movie *Wall-E,* indolent, obese humans are waited on by machines that seem both more industrious and more intelligent than them. How plausible is this scenario? If the machines are truly autonomous then they are capable of coming up with their own goals. Some of these goals may require harming humans. I suspect that there's a good chance that the machines will justify this harm in pursuit of goals that they deem very important even if we've programmed them to feel warm fuzzies when benefiting humans and cold pricklies when harming them.

There's another problem. I've argued that it would be prudentially irrational to upload if doing so involves destroying your biological brain and body. Artificial intelligences may feel differently about such acts. There are many acts that seem unfriendly to less intelligent beings but actually are not. For example, taking your pet to a vet involves acts that seem unfriendly to the pet. Painful injections that protect against potentially fatal illnesses are one example. We recognize these injections as benefiting rather than harming our pets. Artificial intelligences may think they have similar justification for uploading us into machines—against our protests. We needn't attribute any malicious intent on the part of these ever more powerful artificial intelligences. They may want to make the benefits of super-intelligence available to MOSHs. Artificial intelligences might view measures that we rightly fear as benefiting us. They are likely to reject Searle's skepticism about their capacity for thought. Even if they really are incapable of conscious thought, they will probably assert such a capacity. For them, uploading may seem to bring many benefits without recognizable costs or, at least, without costs recognized by them. And once MOSHs have been uploaded their biological residue can be freely absorbed and computationally optimized.

A Human-Friendly AI Research Program?

In science fiction stories, space navigators diligently keep their distance from the event horizons of black holes and so avoid being sucked into them. Is it possible for us to avoid the Singularity's event horizon?

One way to do so might be to stop, or at least dramatically slow, research in AI. Kurzweil is doubtful about the viability of this strategy. He says: "Absent a worldwide totalitarian state, the economic and other forces underlying technological progress will only grow with ongoing advances."[15] Technological advance is promoted by the recognition of "overwhelming benefits to human health, wealth, expression, creativity, and knowledge."[16] This is why any social arrangement that respects individual choices is bound to see the development of these technologies.

I think Kurzweil underestimates the prospects for the successful regulation of AI. We should beware of overly simplistic assertions about the pointlessness of standing in the way of technological advance. Nick Bostrom makes a helpful suggestion about the power of regulators to influence "the *rate* of development of various technologies and potentially the *sequence* in which feasible technologies are developed and implemented."[17] He encourages a focus on *differential technological development*, which he presents as "trying to retard the implementation of dangerous technologies and accelerate implementation of beneficial technologies, especially those that ameliorate the hazards posed by other technologies." Bostrom gives an example from biotechnology: "[W]e should seek to promote research into vaccines, anti-bacterial and anti-viral drugs, protective gear, sensors and diagnostics, and to delay as much as possible the development (and proliferation) of biological warfare agents and their vectors."[18]

Bostrom's chief focus is on the differential development of distinct potentially harmful and protective technologies. The cases that interest me do not straightforwardly fit his model. I'm interested in cases in which the existence of a nonharmful application of a particular kind of research may serve to lure researchers away from the potentially harmful applications of the same basic science. Though human curiosity may make a global ban on the technological applications of a particularly fascinating basic science unsustainable, a nonharmful application may attract the incorrigibly curious away from potentially harmful applications.

For example, we can distinguish between the uses of nuclear fission to make power stations on the one hand and to make immensely destructive weapons on the other. If it's as environmentally friendly as some believe, the former use of nuclear fission may—to use Kurzweil's words—promote "human health, wealth, expression, creativity, and knowledge." The latter seems pretty certain not to. Suppose we have a fundamental need to both better understand nuclear fission and make use of it. If researchers can be directed toward the beneficial uses of a basic science and away from its harmful applications, then there's some hope that liberal democracies can facilitate their citizens' access to nuclear power while preventing them from deploying nuclear weapons. I think there's a similar distinction to be made in AI. Some directions in AI research seem clearly congruent with "human health, wealth, expression, creativity, and knowledge." We can expect there to be democratic support for their development. Others seem not to be. The citizens of liberal democracies should be generally supportive of restrictions on these applications of AI.

How should we distinguish human-friendly from human-unfriendly AI? I propose to separate the two kinds of research based on a view of the mind's structure that comes from philosopher and cognitive scientist Jerry Fodor. Fodor argues that some of the mind's functions are performed by modules.[19] Many of these modules are located on the mind's peripheries and have the job of mediating its interactions with the external world. Examples of modules are those involved in the recognition of faces, the perception of colors, and the processing of auditory information. A module may direct the movement of our limbs as we attempt to catch a ball.

Fodor presents modules as having some or all of the following properties. They are innate rather than acquired. Whereas we may learn how to distinguish male from female chickens and Toyota Corollas from Toyota Priuses, the low-level visual processing that permits these visual discriminations are not learned—we don't learn to see dark colored surfaces as different from light colored surfaces. Modules are domain specific, meaning that they are designed to respond only to specific categories of input. For example, the color perception module responds only to surface reflectance properties, and not to other varieties of perceptual information. The language module is triggered by spoken language, and not by the noise of a jack hammer. The operation of modules is autonomous of other mental operations, which explains why optical illusions continue to appear as illusions even after they've been explained to us. One of the lines in the Müller–Lyer illusion continues to seem longer than the other even after we've seen the arrows removed and replaced, conclusively demonstrating that they're the same length.

Fodor doesn't think that the mind is composed entirely of modules. The high-level cognitive processes that perform analogical reasoning and fix belief do not possess the properties listed in the previous paragraph and are therefore not modular. What Fodor calls the "central processes" form beliefs in response to the outputs of modules and transmit information to other modules responsible for directing behavior.

Fodor's view of the mind's architecture is one possible source of guidance on how to distinguish between AI applications that promote human welfare and those that do not. We could restrict AI's encroachment on our minds to modular functions. This distinction seems to be in line with the advances that have already been achieved. Cochlear implants are essentially replacements for damaged parts of auditory modules. Consider Kurzweil's work on computers capable of understanding speech and reading written language. These are modular functions. A neuroprosthesis that preformed them inside the head of a deaf or blind person would be assisting a module. The same points apply to enhancements. An implant that permitted you to perceive ultraviolet light would be an enhancement of the human color-perception module.

Fodor presents modules as performing a kind of low-level, nonconscious thinking. If this is so, then we should expect Searle to identify a significant gap between the modules and any neuroprostheses that replace them. The replacement of a mental module by a computational device would excise from your mind a category of thoughts, replacing them with insensate, mindless programs. That may be some kind of loss. But how great a loss is it? We're not directly aware of the operation of modules in the first place. It's the results of their processing that present to conscious awareness. Their operations are therefore unlikely to contribute much to our senses of ourselves. Even followers of Searle are likely to accept that the really important thing about the module is that it delivers accurate and useful information to your central processes.

There's a further plus. If AI were to be restricted to modular functions then there's little threat of a *Terminator*-style take over by computers. Suppose a collection of neuroprostheses performing a variety of modular functions were somehow to find themselves attached to one another. The neuroprostheses may individually dramatically outperform their human equivalents. But autonomy is a function of the central processes. This agglomeration of computerized modules is unlikely to have any capacity to act independently of its creators. It's unlikely to take it upon itself to trigger a global nuclear war.

So, which research paths in AI would we deem the equivalents of nuclear physicists working on shoulder-launched nuclear bombs? We should be wary of the plans of AI researchers to electronically upgrade any mental function that we're consciously aware of. Neuroprostheses that help you, say, decide what anniversary present to buy, make your memories of your wedding day more phenomenologically vivid and exciting, or enable you to better understand obscure French philosophy would be taking on some of the functions of what Fodor calls the central processes. Those who suspect that Searle could be right should fear that computerization of these kinds of mental functions will infringe on our conscious experience. Also, once these mental functions are successfully computerized, they could give rise to autonomous, and possibly unfriendly, artificial intelligences.

This recommendation about how to draw the line between acceptable and unacceptable work in artificial intelligence is contingent on a certain view of the structure of the human mind. We shouldn't bank everything on the truth of Fodor's view. But I think it shows that there could be a nonarbitrary, democratically viable way to distinguish human-friendly from human-unfriendly AI.

5 The Therapist—Aubrey de Grey's Strategies for Engineered Negligible Senescence

Shakespeare chose a comedy to present one of the most unpleasant facts about human existence. In *As You Like It* he has Jacques describe life as a play made up of seven acts. The first act casts us in the role of the infant, "mewling and puking in the nurse's arms." We progress to whining school-boys (or schoolgirls) with our satchels, and graduate to the roles of lover and soldier. Life's seventh and final act is a "second childishness" followed by "mere oblivion, / Sans teeth, sans eyes, sans taste, sans everything." Perhaps most depressing of all—this is a best-case scenario. You get to perform the seven-act play only if you aren't struck down by smallpox or a cannon shell sometime in the fourth act.

Aubrey de Grey is a gerontologist based in Cambridge who wants to rewrite life's play. His impressive beard makes him look a bit like Rasputin. The good news is that de Grey's eyes convey some of the Russian monk's intensity but none of the malevolence. To end human aging, he offers *strategies for engineered negligible senescence*, or SENS. The very long final act of de Grey's new play of life will be an indefinitely long period of (very cautious) soldiering and loving interspersed with brief periods of relaxation.

This is not the first time we have come across the idea of life extension. Kurzweil presents a machine-mind that will colonize the universe as the final consequence of the law of accelerating returns. This machine-mind will live precisely as long as it wants to. Kurzweil does take a keen interest in the states of his biological brain and body.[1] But for him, extending biological existence seems to be more of a delaying tactic whose purpose is to keep us around until the advent of machines that will become more secure vehicles of our identities. In chapter 4 I undermined some of the appeal of this version of immortality. De Grey's vision of life extension is altogether less alienating than Kurzweil's. He doesn't view biological bodies as disposable escape pods that will have done their job if they keep us alive just long enough for us to achieve electronic immortality. According to de Grey, we don't have to turn into human–machine hybrids like the Borg of *Star Trek* or the Cybermen from the BBC TV series *Doctor Who* to live for thousands of years.

Ending human aging will cost monumental sums of money and require armies of researchers. De Grey's initial demand is for one billion US dollars.[2] He's confident he can use this money to demonstrate SENS's viability. The final price tag will be dramatically higher, dwarfing the $19,408,134,000 that it cost to send men to the moon.[3] It's likely to be the most expensive cooperative endeavor ever attempted by humans. De Grey is sure that the money is worth spending. He rates aging as "humanity's worst problem."[4] Since SENS is the only systematic response to aging, it must be one of our top moral priorities.

SENS requires major breakthroughs in medical science. The good news is that we don't have to await final victory over aging to be the recipients of millennial life spans. De Grey's immediate goal is *longevity escape velocity* (or LEV), which will be achieved once anti-aging therapies consistently add years to our life expectancies faster than age consumes them. LEV will come well before SENS's completion. Anyone who is alive when anti-aging research reaches LEV and has access to the full range of anti-aging therapies has a good chance of going on to celebrate his or her own personal millennium. De Grey proposes that if SENS receives adequate funding, there is a fifty–fifty chance that we will achieve LEV within "twenty-five to thirty years."[5] De Grey's birth year of 1963 explains his sense of urgency. He clearly hopes that his vigorous proselytizing and frequent pleas for cash will tip the odds in his favor. Appeals for cash are both preface and post-script to all of de Grey's popular presentations. His "show me the money" approach to public speaking has led him to make frequent addresses to the staff and managers of Google Corporation, a company that in mid-2007 had a market capitalization of US $149.2 billion. De Grey's implied question is why that money is being used to improve an Internet search engine rather than being given to him so that he can develop SENS.

Of all the things that defenders of radical enhancement want for us and our descendants, life extension seems to be the most straightforwardly appealing. We commit ourselves to punishing exercise regimens and forgo delicious-looking slices of cheesecake in the hope that these burdens and sacrifices will help us to live just a bit longer. The therapies that may emerge from SENS differ from these established methods by potentially adding to our lives not months or years, but centuries or millennia. And if we accept de Grey as a self-advertisement for SENS, they may be perfectly compatible with the lack of a physical fitness program and the consumption of large quantities of beer and potato chips.

This chapter describes some of the science behind SENS. I argue that de Grey has said enough about how we might put an end to aging to be taken seriously. The reach of SENS clearly exceeds the grasp of early twenty-first-century medical science. It belongs in the same category as putting human colonies on Mars and cloning woolly mammoths—things that we might be able to do eventually if we try hard enough. The question of whether SENS warrants this effort will be addressed in chapter 6.

Strategies for Engineered Negligible Senescence

The search for a fountain capable of restoring the youth of anyone who drank from it is usually presented as a metaphor for misguided endeavor. Many scientists think Aubrey de Grey is every bit as misguided as Ponce de Leon, the conquistador who scoured Florida for the mythical fountain. At least some of the opposition to de Grey's science is provoked by de Grey himself. He's quite a maverick. De Grey was employed at Cambridge University as a computer scientist providing support for a genetics lab. It seemed to him that not enough was being done about one of the most horrible aspects of human existence—death and disability caused by aging—and that he was the man to do something about it. This decision seemed to be vindicated when he proceeded to get papers published in the top journals in the field of aging research. Although de Grey's career path is an unorthodox one in contemporary gerontology, he argues that his training in computer science is actually an advantage: It has instilled in him an engineer's approach to problems. De Grey's motive for investigating aging is not the disinterested pursuit of knowledge claimed by many scientists in academia; he views theories about aging as valuable only insofar as they contribute to ending it.

De Grey's engineer's account presents aging as damage that accumulates in our bodies as a by-product of being alive. Living wears out parts of our bodies in the same way that driving wears out parts of our cars. More fully, aging is "a collection of cumulative changes to the molecular and cellular structure of the adult organism, which result from essential metabolic processes, but which also, once they progress far enough, increasingly disrupt metabolism, resulting in pathology and death."[6] The damage builds up in an insidious way. A forty-year-old may be approximately as intellectually and physically capable as a twenty-year-old. But twenty years of accumulating damage is beginning to affect the forty-year-old in much the same way that long-term infestation with borer beetles affects a wooden house. Though the house may look much as it did before the beetles got to work, each day of continued infestation brings catastrophic structural failure nearer. So it is with age-related changes. With each day lived, damage accumulates and the likelihood of catastrophic failure increases.

It would be great if we could prevent age-related damage from occurring in the first place. De Grey doesn't think this will be possible if we are to get on with our lives rather than freezing ourselves in liquid nitrogen. The aim of SENS is therefore to repair the damage sufficiently well and often that it does not accumulate. Success will make us negligibly senescent, a state we will have achieved when certain depressing facts about human life cease to apply. Currently, eighty-year-olds celebrating New Year's Eve are considerably less likely to be around for the following year's celebrations than are twenty-year-old fellow revelers. Both sets of revelers are equally imperiled by meteorite strikes, jumbo jet crashes, and other exceedingly violent causes of death originating from outside of them. But the continued existence of the eighty-year-olds is under greater threat from the accumulated changes that come about with aging. Their brains and bodies are more likely to experience some catastrophic failure incompatible with survival.

Negligibly senescent eighty-year-olds will have the same very low probability of dying in the coming year as negligibly senescent twenty-year-olds. Although he is sometimes labeled an immortalist, de Grey does not offer immortality. He explains that "Immortality means inability to die, i.e., a certainty of never dying. . . . [T]here is always a non-zero probability of dying some time—and indeed a non-zero probability of dying in any given year. So . . . we will never make ourselves immortal."[7] SENS has no answers to misdirected buses and other killers that originate from outside of our bodies. De Grey thinks that if we repair the damage well enough then our life expectancy at birth could be approximately one thousand years. One thousand is just an average. Some negligibly senescent people will have fatal encounters with meteorites aged twenty whereas others will live until twenty thousand. People who aspire to become negligibly senescent would do well to dispense with any romantic visions of tearful death-bed farewells to loved ones. Death for the negligibly senescent is pretty much guaranteed to come unheralded.

So how does de Grey propose to achieve this? If you present your desire for life extension to your local doctor you're likely to hear something cribbed from the ancient Greek philosopher Aristotle, namely, that it is best to live moderately. Moderate exercise sheds calories and improves cardiac health. But excessive exercise promotes DNA-damaging free radicals and risks cardiac arrest. Moderate quantities of alcohol shift fat deposits from the arteries. Excessive quantities shrink the brain and destroy the liver. The recent fad for calorie restriction represents one departure from Aristotle's prescriptions. People practicing this method of extending life limit themselves to a diet that contains 20 percent to 40 percent fewer calories than what has been considered a nutritionally adequate diet. This method appears to have extended the lives of organisms ranging from fruit flies to squirrel monkeys. According to the opponents of calorie restriction, there turns out to be no such thing as a therapeutically free missed lunch. People practicing this method of extending life find their capacity for intense exercise reduced and their desire for sex diminished. Furthermore, calorie restriction has not been conclusively demonstrated to work in humans.

De Grey proclaims himself uninterested in approaches such as calorie restriction, which don't really get to the core of the problem—and would be hard to reconcile with his love for beer and potato chips. His preferred remedies address the "fulcrum of aging."

This focus on the fulcrum of aging won't just extend our lives. In a talk de Grey gave in 2006 at the Technology, Entertainment, Design (TED) conference he flashed up a power-point slide headed by the question "Why the doubt?"[8] The slide contained two pictures, one of children running around playing soccer, the other of a couple of elderly matrons who seem to be immobilized by age. The caption for the former picture is "fun" and for the latter "not fun." De Grey helpfully explains that "getting frail, and miserable, and dependent is not fun." The things that make us more likely to die as we age also take much of the pleasure out of being alive. Herein lies one of the main failings of current medical therapies and technologies. Chemotherapies and portable oxygen cylinders often extend the lives of people with cancer or emphysema in ways that make their remaining months or years miserable. By contrast, if all goes according to plan, the years that SENS provides will overwhelmingly be healthy, youthful ones.[9]

Seven Deadly Things

The inquiries of traditional doctors are typically directed at entire organs. They look for problems with a patient's kidneys, liver, brain, and so on. De Grey's view is that if we want to describe all the things that can go wrong with a human being our investigations must descend to the level of the cell. Each of the seven varieties of damage that de Grey calls aging is either a problem with cells or with the way cells relate to one another. These cellular and intercellular mishaps are "the seven deadly things."[10] De Grey claims that this seven-item list captures all of the causes of aging.

The first of de Grey's deadly things is the loss of cells that perform important tasks.[11] Over time, our hearts and muscles atrophy as many of the cells constituting them die and are not replaced. Age-related disease kills brain cells at a rate that can dramatically outpace replacement by neural stem cells. These are especially bad places to lose cells. The death of brain cells impairs mental alertness and memory, and atrophied hearts pump less efficiently.

At the same time as we are losing cells that we need, we are gaining cells that we do not want. A second deadly thing is the accumulation of the wrong kinds of cells in some parts of our bodies.[12] Among unwanted cells are, not surprisingly, certain kinds of fat cells. De Grey is particularly worried about "visceral" fat, which resides within the abdominal cavity and accumulates to levels at which it progressively decreases our ability to respond to nutrients in the food we consume. One possible consequence is diabetes. Expanding bellies aren't the only manifestation of this deadly thing. One reason we become more vulnerable to infection as we age is that our immune systems are increasingly committed to maintaining populations of ineffective immune cells that would ideally be terminated to make way for new cells.

A third deadly thing is mutations to our DNA or to the structure of proteins that regulate gene expression.[13] Mutations are changes to our genetic material that can occur whenever a cell divides and its DNA must be copied, or as effects of environmental agents such as sunlight or nuclear radiation. Cells have sophisticated proofreading mechanisms that strive to fix these mistakes. But the cellular complement of approximately three thousand million pairs of letters of DNA makes some mistakes inevitable. The single worst thing that can happen to a cell is that it acquires mutations that make it cancerous. Some mutations make a cell less efficient. Others may kill it. But in cancer, mutations to a single cell among the trillions that constitute our bodies can lead it to divide without limit and kill us.

Mutations to mitochondrial DNA are de Grey's fourth deadly thing.[14] Mitochondria are components of cells that exist outside of the nucleus and make the energy that powers cells. They have their own DNA, which is especially prone to mutation because it is unsupervised by the DNA repair machinery in the nucleus.

De Grey's fifth and sixth causes of aging are the accumulation in the body of various kinds of waste.[15] Our bodies get rid of some waste by excreting it. But other waste is more difficult to purge. Intracellular waste resistant to the action of special enzymes whose job is to break it down accumulates within our cells. Cell division has the beneficial effect of sharing out and therefore diluting waste. But waste tends to build up in cells that divide slowly or not at all, such as those in the brain, heart, eye, and key components of the immune system. Waste also accumulates outside of cells. For example, in Alzheimer's disease bundles of proteins, called amyloid plaques, build up in the brains of sufferers, displacing healthy brain tissue.

Proteins that form what are known as extracellular crosslinks are a special kind of extracellular junk. They are de Grey's seventh deadly thing.[16] Crosslinking proteins are a particular problem for our arteries. For example, they make artery walls less elastic, leading to high blood pressure.

The Seven Deadly Things Meet More Than Seven Life-Extending Therapies

There is no fountain whose waters will prevent and fix all of these seven types of damage. De Grey thinks we will need a wide variety of therapies. He has ideas, some of them speculative, others closer to realization, about how to address each of the seven deadly things. De Grey himself stands prepared to play a role that is part sergeant major, berating scientists for not seeing how their particular training might be refocused on age-related damage, and part choreographer, helping those already addressing relevant issues to see themselves as part of the grand War on Aging (predictably, given de Grey's penchant for acronyms, the WOA).

One area in which researchers are already on the right track is stem cell therapy. Stem cells are the biggest story in contemporary medical research. They are best conceptualized as cellular blanks. Some stem cells, those taken from early embryos, are sufficiently blank to turn into any of the different cell types constituting a human body, and therefore to replace any kind of cell damaged or killed by disease. Prominent among the diseases that stem cell therapy may cure are diabetes, Parkinson's, and heart disease. Progress in stem cell research is currently hampered by moral concerns about the human embryos that must be destroyed to provide the maximally blank variety of stem cell. In 2007 a scientific advance involving what are known as induced pluripotent stem cells, or iPS cells, seemed to make stem cells less morally vexatious. iPS cells result from adult cells that have had their genes manipulated to make them like cells taken from embryos. So we may have cells with the therapeutic versatility of embryonic stem cells without having to kill any embryos. The massive public interest in stem cell therapy has made large quantities of money available for research in this area almost regardless of moral qualms. De Grey is happy to play the role of choreographer rather than sergeant major here. In his view, while stem cell researchers are doing the right kinds of things, it would be good if they recognized themselves as doing more than fighting a series of separate engagements against a disparate collection of diseases. They should acknowledge themselves as participants in the WOA, learning how to reverse the atrophy to brains, hearts, and muscles that increasingly afflicts each adult human with every passing year.

The following discussion focuses on two areas in which de Grey feels obligated to play the role of sergeant major. The first area is mutations of DNA that lead to cancer. The second is accumulations of junk in cells that lead to Alzheimer's and atherosclerosis, among other diseases. De Grey's sergeant major act is necessary not because the medical establishment is oblivious of the problem posed by these diseases. They currently attract vast sums of money and many talented, highly motivated researchers. From de Grey's perspective, however, much of this money and effort is misdirected. The therapies that it might produce simply aren't good enough for people who want to live indefinitely.

There's a difference between conventional medicine's interest in disease and the SENS perspective. Modern medicine aspires to protect people from the symptoms of cancer, Alzheimer's, and atherosclerosis up until the end of what we view as a normal human life span. This degree of protection is inadequate for people aspiring to negligible senescence who require protection that lasts up to a thousand years and beyond. I will call therapies that meet the higher standard *SENS-able therapies*. The question then becomes, what could count as SENS-able therapies for cancer, Alzheimer's, and atherosclerosis?

A SENS-able Cancer Therapy

Most current approaches to cancer fall into one of two categories. Some researchers are trying to work out ways to reduce our likelihood of getting cancer. They might do this by identifying carcinogens in our environments and working out ways to avoid them. Other researchers are looking for new ways to treat people already diagnosed with cancer.

Neither of these traditional strategies is likely to produce SENS-able therapies. The reason has to do with the rather grim mathematics of cancer risk. According to figures cited by the American Cancer Society, the likelihood of contracting invasive cancer between birth and the age 39 is 1 in 70 for males and 1 in 48 for females. By ages 70 and over, those odds have deteriorated to 1 in 3 for males and to 1 in 4 for females.[17] It's no accident that cancer is more prevalent in the old than it is in the young. The cells of an older person have had a greater opportunity to have their DNA miscopied or to sustain damage in ways that make them cancerous. The transmission by cell division of a genetic error in a single parent cell to two daughter cells means that cancer risk increases exponentially. We know from our investigation of Kurzweil's law of accelerating returns that although exponential increase may take a while to really get going, once truly underway its effects can be especially dramatic. There's an obvious implication for people who plan to live for hundreds of years. The risk of cancer will increase exponentially with each year that SENS extends human lives. As you approach your 301st year, it is doubtful that even the most studious avoidance of carcinogens will compensate for your increasing risk of cancer. A three-hundred-year-old is likely to find that her every waking moment is devoted to taking chemotherapies for each of the multitude of cancers that is beginning to afflict her. She won't be happy with the way many current cancer drugs work. They aim to prolong life by slowing the growth of a cancer. This might be good enough for a fifty-year-old who hopes to live until sixty, but it certainly won't be good enough for a three-hundred-year-old hoping to live for another thousand years. Even the slowest growing tumor is likely to grow fast enough to cancel her next millennium.

The summer 2009 issue of the popular science magazine *Discover* offered a demonstration of the divergent cancer agendas of conventional medicine, on the one hand, and SENS, on the other. An excerpt on cancer from de Grey and Michael Rae's book, *Ending Aging*, appeared in an article entitled "From Here to Eternity."[18] This was followed by an interview with Dartmouth Institute for Health Policy and Clinical Practice professor H. Gilbert Welch, who defended the view that cancer is systematically over-diagnosed and overtreated.[19] According to Welch, we impose physical, mental, and economic consequences of cancer therapy on "millions of patients who would have done just fine without treatment."[20] Many cells that satisfy the current pathological definition of cancer either never progress or progress so slowly that a patient dies of other causes. These benign varieties of cancer are ubiquitous—there's some evidence for the view that they vastly outnumber the killers. Welch cites a study that indicated that, on average, fifty men had to be diagnosed of and treated for prostate cancer so that one among them could benefit.[21] Forty-nine of the fifty would undergo painful and debilitating treatment for tumors that neither shorten lives nor produce unpleasant symptoms. Welch wants diagnostic tools better able to sort benign from malignant cancers.

Welch's observations are intended for an audience whose members hope to achieve some age in the normal range for humans. A test that told you that you had cancer but that it would cause you difficulties only if you lived to be one hundred and fifty would be good news for those who hope to live long enough to see their grandchildren grow up. But it's bad news for aspiring millenarians.

De Grey's SENS-able solution to cancer is WILT, which stands for *whole-body interdiction of lengthening of telomeres*.[22] He admits that WILT is very speculative and that it will almost certainly not be among the first wave of life-extension therapies. He concedes that "when they first hear about it, virtually all [his] colleagues think the WILT proposal is utterly mad."[23]

Reduced to its essentials, WILT involves excising from all of our cells the gene that enables the production of an enzyme known as telomerase. On the face of it this seems like a puzzling thing to want to do, especially given the things that have been said about telomerase since it was first identified in 1984. Since then people have speculated about a possible link between telomerase and human life spans. But they tended to think that the chief problem for people wanting to live longer was that we did not have enough of the enzyme. The plan was to give us more telomerase, rather than to do as de Grey wants, and eliminate it altogether.

To understand these contrasting plans we need to know something about what telomerase does. Telomerase extends regions at the end of chromosomes known as telomeres, noncoding stretches of DNA whose job appears to be to protect the DNA that actually makes proteins. It so happens that each time a cell divides, its telomeres get shorter, making them progressively less able to prevent damage to the coding DNA. Thus, the length of their telomeres sets a limit on the number of times a cell can divide. This is the *Hayflick limit*, named for Leonard Hayflick, the biologist who originally hypothesized it. Telomerase's capacity to lengthen telomeres explains why people looking to extend human life spans thought it would be good if we could give our bodies more of it. Telomeres that stay long, rather then getting shorter, offer better protection to our coding DNA. The cell lines composing our brains and bodies could continue indefinitely, thus extending our lives.

It turns out that almost all of our cells possess the gene for making telomerase. In the vast majority of them this gene is switched off, rendered unable to produce the enzyme. Only a few of our cells have active telomerase genes. This is a bit of a mystery. Why should almost all of our cells possess the machinery necessary to make this rejuvenating enzyme yet mostly not bother to turn it on? De Grey thinks he knows the answer to this question. The vast majority of our cells turn their telomerase gene off as a defense against cancer. Cancers kill us by dividing without limit. A tumor whose cells lack telomerase, or some other way to continue to grow, simply can't get very big. Its cells will begin to divide in an unruly fashion but will then just stop when they exhaust their telomeres. If telomerase were readily available it would be much easier for cancers to get big enough to kill us.

However, disabling the telomerase gene turns out to be a far from perfect defense against the enzyme's ill-effects. The cells that comprise a cancer are unstable. The changes that permit them to begin dividing beyond their proper limit pave the way for a variety of other genetic mutations. Among the things subject to random change are the proteins that regulate gene expression. A cell that happens upon a change that unshackles the telomerase gene is free to continue to divide when the surrounding cancer cells have run out of possible divisions. It thus becomes more prevalent, enabling the cancer to grow to a point at which it kills us.

WILT aims to prevent cancers from turning the telomerase gene on by entirely removing it. A cancer whose cells lack the telomerase gene, or some other way to grow without limit, is very unlikely to be lethal. Although it's not so unlikely that at least one among the many millions of cells that make up a tumor can accidentally acquire the ability to switch the telomerase gene back on, it is incredibly unlikely that any of these cells could invent one from scratch. It took millions of years of evolution by natural selection to build the telomerase gene in the first place. Cancers just don't have enough time to reproduce this creative achievement.

De Grey suggests we might remove the telomerase gene from every cell in our bodies by means of genetically modified viruses. Viruses frequently alter the genetic material of organisms that they infect. De Grey hopes that one can be redesigned to systematically excise the telomerase gene. The viruses that medical researchers have trialed are hit-or-miss. They make appropriate alterations to the DNA of some target cells, but miss others. This is a problem, because any cell that retains its telomerase gene might give rise to a lethal cancer. De Grey is confident that scientists will soon (or, at least, eventually) discover more efficient techniques of genetic modification.

What effects might this change have beyond dramatically reducing our likelihood of getting cancer? Most of our cells have switched their copy of the telomerase gene off, so removing it would make no difference to them. But there are some cells that require an active telomerase gene because the telomeres that they start with are simply not long enough. The stem cells that make blood are examples. Excising the telomerase gene from them would soon deprive our bodies of the ability to make new red blood cells, white blood cells, and platelets, killing us pretty promptly. Similar problems would arise in respect of the stem cells in the gut, lung, and skin. De Grey proposes that we use stem cell technologies to provide the necessary replacements. Our bodies would be periodically infused with blood, gut, lung, and skin stem cells that have long telomeres but no telomerase gene. De Grey estimates that this replacement process would have to happen approximately every ten years.

Getting rid of telomerase isn't the whole story. There are cancers that manage to lengthen their telomeres without using telomerase. Needless to say, de Grey has suggestions about how to prevent them.[24]

What are WILT's prospects? Clearly WILT is SENS at its most speculative. This said, I'll briefly explore two problems for de Grey's plan to make cancer impossible.

First, I wonder if removal of the telomerase gene really would give millenarians the degree of protection from cancer that de Grey supposes. If he's right, then purging the body of telomerase may impose strict limits on the growth of tumors. A tumor whose growth is governed by its cells' nonrenewable telomeres may not get big enough to kill us—by itself. But WILT does little to reduce the *frequency* with which human cells enter the first stages of cancer. Cancer risk increases quite significantly over the course of a normal human life. If, as seems likely, this rate of increase is not only exponential but continues beyond the normal human life span, then the early stages of cancer could be very common occurrences for multiple centenarians. A single growth may be subject to the discipline of its telomeres and therefore won't be fatal, but a cluster of such growths could be a different matter. WILT itself does nothing to stop early cancers from accumulating to the point that they begin to interfere with some vital system. A group of such growths on the brain stem could be just as lethal as a single growth using telomerase to grow without limit.

A second problem for WILT lies in its dependence on the truth of de Grey's view of the relationship between telomerase and longevity. A 2008 dispatch from the front line of our species' war against disease further complicated our picture of this relationship. It seemed to support the earlier intuition that increasing our bodies' supply of telomerase might rejuvenate us.[25] Rita Effros, an immunologist based at the University of California, Los Angeles, boosted the ability of immune cells to fight viral infections by inserting the key element of the telomerase gene into them. Similar results were achieved without genetic modification by means of a drug, TAT2, that accelerated the production of the enzyme. Human immune function does tend to decline as we age, so TAT2 does seem properly conceived as rejuvenating. Boosting telomerase production seemed to have other rejuvenating effects. Mice genetically modified to produce more telomerase not only lived 50 percent longer, they had less subcutaneous fat, healthier skin, better neuromuscular coordination, and better responses to glucose.

What of the risk of cancer? The *New Scientist* report describes an experiment in which TAT2 was added to tumor cells and appeared not to increase their production of telomerase. Effros is optimistic. She says, "We are fairly confident at this point that TAT2 won't enhance cancer development."[26] This seems just a tad premature. It's one thing for TAT2 not to accelerate the growth of tumors that have evolved a specific way to make use of telomerase. The bodies in which they evolved did not contain TAT2. We might expect tumors that evolve in bodies whose immune functions have been enhanced by periodic injections of TAT2 to be completely different. What is likely is that any tumor cells that happen upon a capacity to use TAT2 will grow very quickly in bodies with copious quantities of it.

De Grey was quizzed about these discoveries. Interestingly, for a man who aspires to completely eliminate telomerase from human bodies, he said that Effros's results were "what we would have hoped."[27] But it seems that if boosting the levels of telomerase has such a wide range of rejuvenating effects, then completely removing it may accelerate aging in an equally varied collection of ways. Obviously these are early days for WILT—not only in respect of the techniques necessary to realize it, but also in terms of the theory behind it.

How WILT Might Help Enforce the Choice between Life Extension and Children

I conclude this discussion of WILT by pointing to one problem that de Grey himself acknowledges.[28] Blood, gut, lung, and skin aren't the only cells that produce telomerase. The germ-line cells in the ovaries and testes responsible for making sperm and eggs produce it too. And it's actually pretty important that they do. Sperm and eggs combine to make new human organisms, and without telomerase each successive generation would begin life with telomeres shorter than those of the preceding one. The human species would soon simply fizzle out. Negligibly senescent people who wanted to have children might need some treatment to, at least temporarily, restore the telomerase gene to their sex cells.

I think there could be a significant upside to this difficulty. It may underline a moral cost of negligible senescence, a cost that should be especially apparent in this era of heightened concern about the impact of our species on the environment. The combination of people not dying, and being fertile for all but the earliest stages of their lives, could lead to a dramatic increase in population, exceeding the planet's capacity to support us. We worry about the enlargement of our carbon footprint represented by transpacific flights and steak meals. If you have an indefinitely long life, you could have indefinitely many children. Someone who has indefinitely many children who themselves have indefinitely many children in effect has an exponentially increasing carbon footprint.

De Grey acknowledges this as a bad potential consequence of SENS, and he doesn't take the intellectually lazy option of positing interstellar travel and the colonization of other planets. Instead he anticipates our making a choice between children and indefinite life spans. De Grey envisages the negligibly senescent renouncing the right to become parents, ceding that privilege to people who continue to senesce. In his imagined future, some people will opt for childless extended lives and others will opt for business as usual, children and three score years and ten. Humanity's net effect on the environment could be unchanged from today.

I wonder how stable this arrangement is. De Grey is confident that there'll be no need to coerce people who've opted of negligible senescence to remain childless. He himself has freely made this choice, something that he considers to be part of the price for his desired indefinite life span. Should we take him at his word? Human lives are already long enough for people to change their minds about having kids. Many people who form a determination in their early twenties never to have kids discover in their mid-thirties that they'd really rather like to. Sometimes this change of mind comes too late. Suppose LEV arrives according to de Grey's timetable.

One thousand years should give him plenty of opportunity to change his mind about children.

Some of de Grey's confidence about the ease with which the negligibly senescent will renounce parenthood comes from the declining birth rates in parts of the world in which women have been empowered. He explains, "Firstly women are finding it more and more possible to occupy themselves in ways that they find more fulfilling than having kids, and secondly they are breaking out of their upbringing that having kids is the one true way to live." He'd like to promote this trend of childlessness by eliminating dolls as toys. De Grey says: "I personally find the bombardment of young girls with dolls and other assorted paraphernalia of motherhood to be every bit as outrageous as the bombardment of young boys with toy guns and other assorted paraphernalia of violence."[29]

I suspect that female emancipation provides less support for de Grey's child-free ideology than he thinks. The way human lives are currently constructed, a woman's most fertile years are also the most valuable for the building of a career. Establishing careers and raising kids are two massively demanding undertakings. Some women do successfully combine them, but many have to choose. SENS could allow them to have both. Many negligibly senescent women will choose to become both captains of industry and mothers. Left to their own devices I suspect that negligibly senescent people may end up having more kids than the two or three that many time-pressed, senescing humans are limited to. It's hard to imagine a better way for healthy, youthful negligibly senescent people to respond to the empty nest syndrome than having more children.

Furthermore, the desire for kids isn't some cultural artifact that can be eliminated by banning dolls. It's an almost universal aspect of human experience. I suspect many who volunteer for de Grey's therapies will pay lip service to negligible senescence's moral costs. Like fecund fifteenth-century Cardinals, they will have children in spite of their vows. But here's the potential upside of WILT. WILT's effects on sex cells may make it harder to overlook its moral cost. Societies that have resolved to enforce a choice between children and indefinite life spans would be able to police access to therapies that temporarily restore telomerase genes to the sex cells of people treated with WILT. Those who forgo WILT may be able to combine extended lives with kids. But unless they find some other SENS-able response to cancer, they'll soon be beset by tumors.

A SENS-able Therapy for Atherosclerosis and Alzheimer's Disease

Atherosclerosis and Alzheimer's diseases are consequences of proteins that accumulate in our cells in spite of enzymes whose job is to break them down. They, like cancer, are the focuses of millions upon millions of dollars of research money. And, as with cancer, most of this money is, from de Grey's perspective, misdirected. It may yield therapies that keep blood pumping through our arteries, or neurons encoding memories well enough to sustain us into our eighth and ninth decades, but it is unlikely to carry us into our eighth or ninth centuries. The problem is that few therapies directly address the intracellular junk that is the root cause of these diseases. They aim instead to mitigate its effects. Thousand-year-olds face the prospect of it accumulating until its effects are beyond mitigation. SENS-able therapies for atherosclerosis and Alzheimer's will therefore have to entirely eliminate the intracellular junk.

What de Grey needs is a therapy that reaches into our cells, locates the undigested junk, and eliminates it, all without interfering with any of the essential cellular machinery. This seems a tall order. De Grey's solution is indicative of a real talent for thinking outside of the box.[30] It involves identifying the therapeutic potential for knowledge that had hitherto seemed completely without medicinal value.

The inspiration for de Grey's solution to intracellular junk comes from something called bioremediation, a strategy for eliminating environmental contaminants such as oil spills. Bioremediation's conceptual starting point is something known as the microbial infallibility hypothesis. This is the idea that any organic energy-rich compound creates a strong selective pressure for nearby microorganisms to evolve so as to degrade it. Essentially, it's the notion that if there's food around some microbe will eventually acquire the ability to eat it. Bioremediation involves using these specially adapted microbes to clean up pollutants. The place to start looking for the relevant microorganisms is wherever the energy-rich compound you want to remove tends to occur naturally. Once found, they can be applied in large doses to the mess. Bacteria that eat hydrocarbons are now routinely applied to oil spills.

If bioremediation is to be of value in cleaning up intracellular junk then we need to find the microbes that have evolved to eat it. De Grey proposes that we look in, of all places, graveyards. We know that microbes capable of doing the job must be there because human bodies decompose. Relatively nutrient-poor bones and teeth hang around for a long time and nutrient-free amalgam fillings and hip replacements for a very long time, but pretty much everything else disappears pretty promptly. It follows that there must be something in graveyard soil that is breaking down the intracellular junk. De Grey has begun the task of collecting microbes from graveyard soil and seeing which ones flourish when introduced into environments that contain only proteins that lead to the cellular malfunctions characteristic of Alzheimer's and atherosclerosis.

De Grey doesn't propose that, once found, we introduce these bacteria directly into the human body. We would probably find them doing to our living bodies exactly what they do to dead ones—skeletonizing them. But he's optimistic that once the specific enzymes that break down the junk are isolated we can turn them into targeted therapies for Alzheimer's and atherosclerosis, and for the age-related damage that shares their causes.

Longevity Escape Velocity

If de Grey is correct in thinking that there are no other causes of aging, then effective therapies for all of the seven deadly things will make us negligibly senescent. They will end aging. Obviously some of the causes of aging are more easily addressed than others. De Grey's suggestions about how to prevent cancer and remove intracellular junk clearly aren't going to provide therapies anytime soon. Indeed it's not hard to imagine them joining a long list of beautiful ideas that fail reality's tests. The good news is that SENS has a failsafe built into it. We can achieve indefinite life spans without being able to repair all aging-related damage. This immunity against failure is provided by the concept of longevity escape velocity (LEV). According to de Grey we will have attained LEV at such time that "mortality rates fall so fast that people's *remaining* (not merely total) life expectancy increases with time."[31] As things currently stand, each year that someone lives decreases the number of additional years it is rational for them to expect. This is what it means to age. LEV will turn this apparent truism into a falsehood. It will arrive when our research programs consistently add more years to life expectancy than the time taken to do the research. Suppose scientists spend one year inventing a therapy that adds two years to the life expectancies of those who receive it. People alive at the beginning of the year who have access to the new therapy should expect more additional life at the end of the year than they expected at its outset. Scientists will have more time to make the therapy even better and to combat the other causes of aging. De Grey envisages a positive feedback loop as the extra years give scientists time to find additional remedies for the seven causes of aging.

At the beginning of this chapter I compared the likely expense of SENS with the expense of the Apollo project. The similarities do not end with their huge price tags. Both require vast, collaborative, multidisciplinary efforts. The concept of LEV sets SENS apart from the Apollo project in one significant respect, however. The Apollo project was intolerant of failure. Had workers on the rocket boosters, life support systems, the heat shielding on the command module, or any other essential system been seriously in error then the mission to the moon would have failed. The intermediate goal of LEV means that lift-off for SENS does not require responses to all of the causes of aging. SENS could succeed even if the majority of de Grey's proposals are wrong. If only a few of them lead to therapies that extend life then we will have more time to come up with better ideas about how to fix the less tractable damage.

De Grey has proposed a timeline for engineering negligible senescence. His intermediate goal is *robust mice rejuvenation* or RMR, which we will have achieved when we take "a cohort of mice of a strain whose normal life expectancy is three years, do nothing to them until they are two years old, and get them to live an average of three more years, i.e., tripling their remaining life expectancy."[32] De Grey guesses that we will be rejuvenating mice in seven to twenty years. These mice will be a significant scientific proof of concept. Almost more importantly, they will transform our perception of death. The rejuvenated mice will shake us out of something de Grey calls the *pro-aging trance*. We mourn deaths from homicide, road accidents, and AIDS because we view them as potentially preventable. But we accept death from aging as a universal and inevitable part of the human condition, something to write poetry about rather than to seriously try to avoid. RMR should persuade us that deaths from aging are preventable and therefore make them intolerable. Any politician who argues against allocating a substantial percentage of tax revenues to SENS will be unelectable. The next milestone is *robust human rejuvenation* or RHR. This will be achieved when we can begin administering therapies to humans in middle age and triple their remaining life expectancy.[33] RHR could follow RMR by fifteen years. In de Grey's imagined worst-case scenario it follows RMR by one hundred years. RHR will get us close to LEV. In the first of a series of talks to the employees of Google Corporation, de Grey explained the relationship between RHR and LEV by saying "The first 1000-year-old is probably less than 20 years younger than the first 150-year-old."[34]

Can He Do It?

So—can SENS accelerate the human species to longevity escape velocity? Or is de Grey a fraud, or at least terribly misguided? His scientist opponents dismiss SENS as "agenda/ideology-driven pseudoscience."[35] In 2005 a journal published by the Massachusetts Institute of Technology, the *Technology Review*, issued a challenge to de Grey's opponents to demonstrate that SENS was "so wrong that it is unworthy of learned debate." A number of scientists undertook this seemingly self-refuting exercise and a panel of experts judged the results, arriving at the unsatisfying verdict that none of the alleged refutations had proven that SENS was unworthy of debate, but neither had its proponents made a compelling case for it. Jason Pontin, the *Technology Review*'s editor, summarized the judges' rationale for this seemingly null judgment as follows: "de Grey's proposals exist in a kind of antechamber of science, where they wait (possibly in vain) for independent verification. SENS does not compel the assent of many knowledgeable scientists; but neither is it demonstrably wrong."[36]

There's something right and useful in the suggestion that SENS "exists in a kind of antechamber of science" and is therefore in some sense outside of it. I want now to explain why the principal focus of next chapter's discussion will be on SENS's status as a moral and social priority, and not on its narrowly scientific status. No one doubts that it is beyond the powers of early twenty-first-century medicine to make humans negligibly senescent. The question is whether we should invest the resources necessary to work out whether and how it can be done.

It's useful to compare de Grey's claim that, given adequate support, we can end aging with US President John F. Kennedy's commitment, made on May 25, 1961, to send humans to the moon. This was impossible given the science and technology of the early 1960s. President Kennedy was issuing a challenge to scientists to do something clearly beyond the science of his day.

Those who opposed what would soon be baptized the Apollo program on scientific grounds would have to do more than demonstrate the inadequacy of the blueprints for a lunar module that existed at the time of Kennedy's proclamation. In the early 1960s these probably existed only as rough sketches. Scientist opponents would have to show that it would be impossible to build one ever, or at least within the decade that Kennedy proposed to put humans on the moon. The leap from the science of the early twenty-first century to the ending of human aging is bigger than that from the science of 1961 to putting humans on the moon. And de Grey has asked for more time. Bearing this in mind, critics of SENS on purely scientific grounds will need to show that indefinite life spans are a scientific impossibility, or that they are so difficult to achieve that we'll only see them in 250 years or 300, and not in the 25 years or 30 suggested by de Grey.

One question concerns who's best placed to decide whether and when we can give humans indefinite life spans. Many gerontologists are pretty confident that it's not de Grey. Those who accepted the *Technology Review* challenge made the point that within gerontology he's an arriviste who hasn't done the bench science that is part of the standard apprenticeship of scientists in gerontology. Some of his scientist opponents conclude that his proposals about human aging and how to end it are therefore not to be taken seriously.

I disagree. I think de Grey is about as well placed as anyone to make these assessments. The judgment of whether or not something can be done and how long it might take to do it requires a variety of expertise that differs from that required to conduct specific relevant research.

Consider the predicament of Kennedy's political advisors trying to work out whom to ask about the feasibility of sending humans to the moon within the decade. They could consult a collection of *specialists*, experts on specific parts of what must be done to get humans to the moon and back again. One specialist might know a great deal about conditions close to the lunar surface. Another might be the world authority on what happens to objects when they enter the Earth's atmosphere. Still another specialist might know all that is currently known about keeping humans alive outside of our planet's atmosphere. These individuals may be essential to a successful moon mission. But they aren't necessarily the best people for Kennedy's advisors to talk to. A successful moon mission must put all of these jigsaw pieces of specialist expertise together in the right kind of way. For example, the shielding that permits returning astronauts to survive reentry into the Earth's atmosphere must not make it impossible to design a successful lunar module. Specialists aren't guaranteed to know how to do this. After all, they may be drawing on vague memories of various *Scientific American* articles about all the other advances that will be required to fly humans to the moon and back. Instead of petitioning specialists, Kennedy's advisors should ask *generalists*, scientists who may be conducting cutting-edge research in no particular field, but who are very well informed about research in all the areas predicted to be relevant. This variety of scientist is better placed to decide whether the various jigsaw pieces of specialist knowledge can be combined to make a successful moon mission. Of course, when it comes to actually building a lunar module they'd need to consult the relevant specialist, rather than a generalist who may be confident that it can be done, and have a fair idea about how it should be done, but be a bit hazy on the details.

De Grey's maniacal dedication to the task of ending aging has made him much better than the world's expert on the *popular* science of aging. His book, *Ending Aging*, serves as testament to the breadth and depth of his knowledge of the science required by SENS. He does seem to be among the people best qualified to express an opinion on whether SENS could work. His declared passion for SENS should probably disqualify him as an impartial observer. And this should make us suspicious of some of his proclamations. But we should be honest about the nature of this suspicion. It is grounded in skepticism about de Grey's partiality, rather than about his scientific expertise.

What then are we to make of de Grey's specific proposals? Is he like the generalist who's just advised President Kennedy that it's worthwhile trying to go to the moon within the decade who then sets about building a lunar module? I think de Grey's various suggestions about how SENS might be achieved are best viewed as working hypotheses. We shouldn't bet everything, or indeed anything, on them being literally true. What is more important than their truth or falsity is their propensity to lead to therapies that really do fix the damage.

WILT is best thought of in this way. Some social conservatives say that since the poor will always be with us we shouldn't bother doing anything to help them. WILT is a response to conservatives in medicine who hold that since cancer will always be with us therapies can only ever aspire to a defensive role, mitigating some fraction of the disease's horribleness. Even if it does not lead to therapies it at least shows that a future in which humanity is free of cancer is not inconceivable. How likely would someone traveling forward in time be to find people being treated with WILT? I suspect not very. There are many contingencies. For example, if stem cell therapy doesn't work then all bets are off. We would also need to discover a safe technique for removing the telomerase gene from every cell in our bodies. Perhaps the biggest contingency concerns the discovery of some more straightforward way of preventing cancer. One eventuality that would be at the same time very bad for WILT and great for SENS would be the invention of nanobots that safely kill all and only cancer cells. If, by 2020, we're introducing these into our bodies, there'll be little incentive to persevere with WILT.

SENS is best challenged as a social priority rather than on purely scientific grounds. The pertinent questions are: Should the relevant experts in human aging heed de Grey's call? Should money that could be spent addressing the problem of climate change, or improving the lot of people in the poor world, instead be spent on SENS? Those who opposed the Apollo program on scientific grounds have been proven wrong—it was possible to send humans to the moon within the decade promised by Kennedy. But the question about the Apollo project as a social or scientific priority remains open. Would the billions that were spent on it have been better spent on unmanned space exploration, building a space station, or perhaps to investigate the possible long-term consequences for the climate of the quickening pace of industrialization in the world of the 1960s? In chapter 6 I address SENS as a moral and social priority.

6 Who Wants to Live Forever?

Who dares to love forever?
When love must die
—"Who Wants to Live Forever," *Queen*, lyrics by Brian May

In the 2000 remake of the 1967 movie *Bedazzled*, the devil, played by Liz Hurley, offers to help Eliot Richards, a romantically ill-fated loser played by Brendan Fraser. She will grant seven wishes in exchange for his immortal soul. Eliot accepts and makes the obvious first wish to be rich and powerful. The devil gives him money and power . . . as a drug lord whose wife despises him and whose business partners are plotting to kill him. The NBA basketball stardom he wishes for comes with a very small penis and an exceedingly low IQ. The devil grants Eliot's desire to be intelligent, witty, and well endowed, but as a gay man who therefore has no chance with the woman of his dreams. And so on. In each case Eliot asks for something that seems to have obvious appeal. There's always a deal-breaking catch that he doesn't anticipate.

If Eliot had been happier with his life he might have wished for much more of it. The devil would have been too shrewd to grant immortality—that would infinitely defer payment of her fee. But negligible senescence might have been on offer. She would have been confident that some fatal mishap would eventually send Eliot's soul her way. Her challenge would have been to find a way to grant a request for negligible senescence but only in a form guaranteed to make it not worth having.

De Grey looks more like traditional representations of Beelzebub than does Hurley, but he's not out to cheat us, and he has absolutely no designs on our immortal souls. Even if de Grey's not on a sabotage mission his infectious enthusiasm for negligible senescence may lead us to make the same mistake as Eliot. In chapter 7 I will argue that the scenarios used to market radical enhancement tend systematically to omit information about its true costs. They rely on a human tendency to subconsciously fill in details in ways that seem intuitively to go with the stated central facts of an imagined situation. Eliot is told that he will be rich and powerful. Stereotypically rich and powerful men have loyal wives and respectful business associates. But money and power grant no immunity from cuckoldry or betrayal. Eliot wouldn't be the only fan to imagine reproductive equipment in proportion with the elongated frames of NBA basketball players. But it's strictly surplus to the requirements of scoring from free throws and making slam dunks.

I suspect that the same is true of de Grey's pitch for SENS.[1] What's true about accepting an offer of negligible senescence is that you're likely to live much longer than you otherwise would. De Grey wants us to think of this as an indefinite continuation of the best bits of our lives as they are or were. For example, he paints this picture of our negligibly senescent futures:

> For the moment, when you retire, you retire forever. We're sorry for old people because they're going downhill. There will be no real moral or sociological requirement to do that. Sure, there is going to be a need for Social Security as a safety net just as there is now. But retirement will be a periodic thing. You'll be a journalist for 40 years or whatever and then you'll be sick of it and you'll retire on your savings or on a state pension, depending on what the system is. So after 20 years, golf will have lost its novelty value, and you'll want to do something else with your life. You'll get more retraining and education, and go and be a rock star for 40 years, and then retire again and so on.[2]

Suppose you don't like golf. Substitute Porsche driving, bridge playing, collecting ancient Roman coins, or any other hobby that you could easily imagine filling your entire life up with. Or suppose rock star isn't your favorite "if only" profession. Substitute movie director, poet, or marine biologist. If this is an accurate depiction of negligible senescence then there seems something perverse in turning it down. Those who do so seem to be saying that there's no activity that they find so enjoyable or important that they wouldn't appreciate more time in which to do it.

In this chapter I argue that the experience of negligible senescence will be altogether stranger than de Grey would have us believe. It's not simply more of the same, where "the same" picks out the best parts of our lives as they are, or were. The decision to become negligibly senescent is likely to transform us into very different kinds of beings with very different kinds of experiences and priorities. It is likely to turn us against many of the activities that currently fill the best parts of our lives.[3]

Deathism and Suicidal Urges

De Grey and his allies have labeled the ideology that opposes negligible senescence "deathism." They present the rejection of an indefinite life span when one is available as tantamount to committing suicide. Mark Walker offers a more circumspect presentation of this point. He thinks that people opposed to suicide should be attracted to indefinite life spans. According to him, there's a similarity between the motive of rejecting negligible senescence and that behind suicide. Walker makes the point that "If we think of suicide as 'voluntarily ending one's own life' then any death that results from refusing to use superlongevity technology looks like an instance of suicide."[4] To play up the reasonableness of this reaction, Walker compares the decision to reject radical life extension with the refusal by someone to accept medical attention following a fall that leaves him with injuries that will end his life in a few days without treatment. According to Walker, "So long as we think of this decision as an act of suicide, it seems that by parallel reasoning we ought to think of the refusal of radical life extension treatment as a form of suicide."[5] The journalist Bryan Appleyard seems to agree. He began his investigation as a confirmed deathist, someone who believed that it is good that we die. But reflection on the topic shook his confidence in the desirability of dying. He writes, "Lying on my deathbed, would I refuse the treatment that would take me back to my late twenties and perfect health? This, surely, is the one offer nobody could refuse."[6]

Actually, the label "deathism" does not capture a genuine difference between indefinite and definite life spans. Unless there's an answer to the big crunch that many cosmologists think will bring the universe and all of its sentient citizens to an end then no indefinitely long life will be infinitely long. Jacques's speech in *As You Like It*, suggesting that all the men and women will still have "their entrances and their exits," will still ring true. Regardless of whether SENS succeeds our lives will be framed by birth and death. The real issue concerns the value of what happens between these events.

It does seem better, in general, to have a longer life than a shorter one. All things being equal, someone who dies at eighty has had a greater number of valuable experiences than someone who dies at age twenty. But this preference for longer rather than shorter may not apply outside of the limits of normal human life spans. It could be rational to desire that the events of birth and death frame an on average shorter but recognizably human life.

To return to Walker's argument, there's likely to be a difference between rejecting treatment for your compound fracture and rejecting de Grey's life-extending therapies. The subject of Walker's example forgoes therapies and so lives out a final few days in intense pain. It seems hard to make a case that a life that ends after a few days of intense pain will contain a greater number of valuable experiences than would the life that you would have had after having treatment for the compound facture. This assessment might be reversed if it turned out that there were something of immense value that you could achieve as you lay dying of your injury. Consider another case in which one makes a decision that predictably shortens one's life at the same time as potentially increasing the value of the experiences it contains.

Those who set life insurance premiums should tell you that the decision to become a mountaineer shortens your life expectancy. Mountaineering is very dangerous. Slightly fewer than one in ten of the climbers who have attempted to reach Everest's summit have perished. And other mountains are even more perilous than Everest. Especially risk-taking types increase the danger by committing themselves to Alpine-style mountaineering, which disdains oxygen-tanks, fixed ropes, and Sherpa guides.

The fact that the life expectancies of mountaineers are shorter than those who make different life choices does not make the decision to become a mountaineer suicidal—or irrational. A mountaineer stranded on Everest is likely to strongly desire to be rescued. Suppose he realizes that no one is coming for him. He may regret his decision to become a mountaineer in the first place, wishing that he'd satisfied his urge for risk-taking by becoming a trader on the New York Stock Exchange. But his regret may not take this form. For truly dedicated climbers it will be limited to the decision to attempt the summit when the weather was treacherous. People who reject de Grey's therapies are also quite unlike people who commit suicide. They want their lives to continue, and they volunteer to have treatment for potentially terminal illnesses. But they don't want to continue in just any way. They want to continue living what they consider to be a recognizably human life. If you think along these lines, when you're told that you have cancer you desperately hope that the chemotherapy works. But you don't wish that you'd been a fundamentally different kind of being, one that could never have contracted cancer in the first place.

It's easy to understand the appeal of mountaineering. You get to go to places visited by few other people on this crowded planet. You have the thrill of scaling sheer cliff faces and detecting invisible crevasses. These experiences make it rationally permissible, though certainly not rationally compulsory, for some people to choose lives that are, on average, shorter than other lives they might have led. In what follows I explore experiences that make it rationally permissible, though not rationally compulsory, to prefer on average shorter existences imposed on humans by aging to the on average longer existences brought by negligible senescence. I predict that when fully informed about both alternatives, most humans will prefer to retain definite life spans even if they are, on average, shorter than indefinite life spans.

There are some ways in which we could increase our life expectancies that most of us would reject. Consider the following illustration of species-relativism about valuable experiences. The Galapagos tortoise (*Geochelone nigra*) has a life expectancy, in its natural habitat, between one hundred and one hundred and fifty years—considerably more than the just over eighty years averaged by rich-world humans. Suppose scientists discovered a way to turn humans into Galapagos tortoises. Volunteers for the procedure would remain conscious throughout the multiple genetic and surgical procedures to ensure the preservation of their identities. The procedure's selling point is an average of twenty to seventy years of additional life. The longer life gives the opportunity for greater numbers of pleasurable experiences. Many of these pleasures will be different from those that populate human lives—the pleasures at locating an especially propitious nesting ground or finding an especially suitable mate are quite different in phenomenological character from their human equivalents. But they'll be every bit as enjoyable by tortoises as are human pleasures by us. I suspect that few humans would volunteer for the procedure. It's a consequence of the species-relativism about valuable experiences described in chapter 1 that we can be too attached to our own distinctively human varieties of pleasure to be persuaded by the argument that longer tortoise lives poten-tially contain greater numbers of pleasures that are, considered individu-ally, every bit as subjectively valuable as human pleasures. Perhaps we can give similar reasons for rejecting the longer life expectancies offered by de Grey.

This example is slightly unfair. The life of a negligibly senescent human is more similar to the one that we currently lead than the life of a Gala-pagos tortoise. Transformation into a tortoise would involve substantial cognitive diminishment with obvious consequences for our values. In what follows I explore the possibility that negligible senescence might threaten experiences that contribute value to human experiences. These experiences give many a good reason to prefer a normal human life span to an indef-initely long one.

Will Negligible Senescence Be Boring?

The most prominent line of skepticism about the quality of radically extended lives comes from the philosopher Bernard Williams. He argues that we would find infinitely or indefinitely long lives boring. This boredom is a reaction to "the poverty of one's relation to the environment," the fact that there are only finitely many ways in which we can interact with the things that make up our world.[7] Fans of golf, Porsches, bridge, or coin collecting may find it difficult to imagine that these obsessions could become tedious. But they haven't had the opportunity to devote hundreds of years to them. The first experiences of the Taj Mahal, the Grand Canyon, bungee jumping, and *Gone with the Wind* are fabulous. Second experiences are still great, but often noticeably less good than the first. This process of becoming slightly less wonderful leads inexorably to their becoming intol-erably boring. The Canadian philosopher Christine Overall proposes that the tedium of a radically extended life might be analogous to the awfulness of severe insomnia.[8] Insomniacs crave the loss of consciousness that comes with sleep. If Overall's analogy is apt then the negligibly senescent may develop an intense desire for the permanent cessation of consciousness brought by death.

It's hard to say what we should make of this fear of boredom. Just how likely are healthy, youthful three-hundred-year-olds to be bored? We can ask insomniacs to describe the unpleasantness of chronic sleeplessness.

There are, in contrast, no negligibly senescent three-hundred-year-olds to consult. The closest approximation was the Frenchwoman, Jeanne Louise Calment, who smoked and drank her way to a life span of 122 years and 164 days. But she and other centenarians may not be particularly reliable guides. The oldest contemporary humans must endure the painful tedium of the aches and pains of aging, things that de Grey plans to abolish. They are more likely to experience frustration at an inability to do the things that they once loved rather than to feel bored at having done them too many times. There seems a big difference between the experience of a present-day centenarian whose mind and body have suffered the ravages of the seven deadly things and that of a negligibly senescent three-hundred-year-old, mentally and physically freer of age-related changes than are today's twenty-year-olds.

The prospect of boredom is more a concern for someone contemplating an offer of immortality than it is for someone deciding whether to accept de Grey's offer of negligible senescence. There's actually a big difference between immortality and negligible senescence. Whereas a negligibly senescent being is *likely* to have a longer life span than a senescing one, an immortal being is *guaranteed* to. Immortal beings have a zero probability of dying over any future period of time. Negligibly senescent beings have, in contrast, a nonzero probability of dying with each year that passes. The difference between them and us as we are now is that this probability does not increase. Though he's sometimes called an immortalist, de Grey's therapies will never reduce the probability of death to zero. SENS offers no protection from out-of-control buses or supernovas. But suppose that an omniscient, omnipotent being were to offer you immortality. You should think very carefully before you accept. An immortal being is not only immune from death that results from accidentally ingesting arsenic, she also cannot die as a result of intentionally ingesting it. That's what a zero probability of dying means. Suppose that Williams turns out to be right, and in the one thousandth year of your immortal existence you begin to succumb to an incrementally and inexorably increasing boredom. By your one-thousand-five-hundredth birthday, every action that is within your power to perform bores you. Your infinite life span will accrue more suffering than the most miserable finite life span.

To return to Overall's analogy of the tedium brought by an extended life with the suffering associated with severe insomnia, beings with indefinite life spans have an option that severe insomniacs might envy. A lethal dose of barbiturates would provide a negligibly senescent being with an immediate and permanent cessation of consciousness. An immortal being could be more like someone suffering from an insomnia that is guaranteed to be beyond treatment.

The difference between negligible senescence and immortality makes the approach recommended by the philosopher John Harris seem warranted. Harris observes, "I would (as of now) be quite happy to sample a few million years and see how it goes."[9] Much in the same spirit, Walker proposes a "superlongevity experiment" to test Williams's hypothesis.[10] If we discover that Williams is wrong, then those testing negligible senescence will have done very well indeed. If life does become intolerably boring then the experiment can be terminated by means of an overdose of barbiturates. Such overdoses are much more likely to come at age 300 than at any age that senescing humans currently reach. And there's actually an upside to Williams's grim forecast. The good thing about eventually becoming bored with an activity that you used to love is the time it takes to become boring. Bridge fanatics are likely to be pretty excited at the prospect of playing enough of their favorite game that they have the opportunity to become thoroughly bored with it, something unimaginable in a mere fifty years of play.

I want to explore a different potential blight on indefinitely long lives—*fear*. Remember that de Grey accuses his opponents of reconciling themselves to death. He thinks that if we free ourselves of the pro-aging trance we will recognize death as something that can and should be avoided and therefore rush to support SENS. I suspect that the fear of death may completely dominate the lives of negligibly senescent people. It will do so to such an extent that it will prevent them from enjoying many of the activities that make our lives pleasurable and meaningful.

Suicide is a sure fix for boredom available to negligible senescent beings but not to immortal ones. This is negligible senescence's advantage over immortality. The great advantage of immortality over negligible senescence is that beings with a zero probability of death need never fear it. This absolute immunity from death opens up a range of exciting activities far too dangerous for us, such as skydiving from the top of the Eiffel Tower *sans* parachute and exploring black holes from the inside. Negligibly senescent beings are likely, in contrast, to develop a risk-aversion that will radically truncate their existences.

The Fear of Driving

In a 2003 interview, de Grey was asked the following somewhat flippant questions: "Why is it that in the year 2003 I still don't have a flying car? When do you think I'll be able to get one?"[11] De Grey might have responded that automotive engineering lay outside of his area of expertise. But he didn't. Instead he answered, "When will they become available: I suspect never, in fact, because quite soon we will know that the end of aging is on the way, and the consequences in terms of increased risk-aversion will be so great that there won't ever be a market for things that risky."[12] De Grey continued that in 1999 he had predicted that "once we cure aging, driving (even on the ground!) will be outlawed as too dangerous for others." He added, reassuringly, "Remember also that when we have so many more years ahead of us, we won't need to be in such a hurry all the time."[13]

De Grey is right to think that negligibly senescent people will be wary of cars. They'll also diligently avoid a variety of other activities that we currently think of as enjoyable or worthwhile. To see why, let's consider how the activity of driving will present to those who are negligibly senescent.

According to research reported by the US Department of Transportation there were, between the years 1999 and 2003, 1.3 fatalities for every hundred million miles traveled in a car either as driver or passenger.[14] This statistic compares unfavorably with travel by bus, train, and airplane. Does this make traveling by car irrational? Perhaps not. But people who've become negligibly senescent or who have achieved longevity escape velocity are considerably more likely to answer this question in the affirmative than we are.

Humans are not entirely rational in our responses to risk. Many people think it's safer to travel long distances by car than by commercial jet liner even though it manifestly is not. We're not particularly rational in the distribution of risk across our life spans. Young people take dangerous drugs and imitate the self-destructive behavior presented in the *Jackass* movies. They've got much more to lose from these activities than do the older people who typically refrain from them. The young should be the ones watching *Days of Our Lives* reruns in the safety of their lounges. Alan Arkin's character in the movie *Little Miss Sunshine*, Grandpa Hoover, is a good advertisement for a more rational approach to dangerous drugs. He takes heroin. When quizzed about his heroin habit he expresses the view that "When you're young, you're crazy to do that shit" but "When you're old you're crazy not to do it."[15] The pleasures brought by the drug are not significantly less than those experienced by addicts in their early twenties. He has less to lose from an accidental overdose (which in fact ends his life). Hormones explain part of the difference in attitudes to risk of today's teenagers and octogenarians. The good news is that de Grey disciples are likely to have attitudes to risk that are more rational than those of teenage boy drag-racers. They are self-consciously reducing and eliminating risks resulting from the seven deadly things. They're also unlikely to go drag-racing in souped-up cars.

One way to approach the rationality of driving is to think systematically about what you stand to gain or to lose through doing it. Presumably when you get into a car you have a reason for going somewhere. Suppose you want to drive to a local cinema to watch a movie. There's a nonzero probability that a car crash will kill you either on the way to the cinema or on the way back. This outcome is very bad. The good news is that the probability of a fatal accident, if you drive sensibly, is extremely low—it's certainly higher than your chance of death from the spontaneous collapse of your house, but still very low indeed. Simplifying somewhat, the expected utility of the trip to the cinema is a very high probability of a brief but enjoyable experience—seeing the movie—minus the minuscule probability of a very bad outcome—death. The fact that a fatal crash is so extremely improbable can make driving to the cinema rational—for a senescing being, that is. Negligibly senescent people should think very differently about seeing the movie. Fatal car crashes are bad for senescing beings, but they're going to be so much worse for the negligibly senescent. A forty-year-old senescing human who gets into a car stands to lose, at most, a few healthy, youthful years and a slightly larger number of years with reduced quality. LEV increases the potential costs of driving by more than one-hundred-fold. At any time that a being who is negligibly senescent dies, an average of one thousand healthy, youthful years are lost. Current road safety TV advertisements do a good job of making senescing types fearfully aware of the disastrous consequences of a momentary lapse in concentration or of an act of recklessness, on your part or on the part of some other motorist. They will inspire even greater levels of fear in negligibly senescent viewers. For the car trip to the cinema to be as uncontroversial a matter for the negligibly senescent person as it is for us, the experience of seeing the movie would have to be over one hundred times better than it is for us. The increasing budgets of Hollywood movies would have to have had commensurate consequences for their quality.

Negligibly senescent people don't have to be terrified into complete inactivity. Instead of braving the highways, negligibly senescent cinephiles could instead just download them onto their computers and watch from the safety of their armor-plated homes. More importantly, SENS may boost our chances of successfully performing activities that really warrant the risk. Remember that SENS does more than just extend lives. It maintains and restores our vitality, something that may increase our ability to do valuable things. De Grey was concerned that the increased risk-aversion of negligibly senescent people might make socially valuable occupations such as fire-fighting extremely unpopular. As the events of September 11, 2001, made patently and painfully clear, fire-fighting is a life-endangering occupation. De Grey related to me his relief on receiving a message from a fire-fighter reassuring him that SENS would in no way weaken his resolve to do his job. According to de Grey's correspondent, "If my body was not aging and becoming damaged beyond repair, I would never leave this job. Most of the men and women I work with consider it an honor to serve others by endangering our lives to protect theirs. So don't think everyone would want to stop being firefighters, because some of us Love it!"[16]

So, on the one hand a negligibly senescent fire-fighter will lose more when she finds her escape from a combusting chemical factory blocked. She'll lose an average of one thousand healthy, youthful years as opposed to perhaps ten or twenty such years and an additional thirty of diminished quality that heroic self-sacrifice takes from fire-fighters in the early years of the twenty-first century. But, on the other hand, as de Grey's fire-fighter correspondent points out, SENS will make them better fire-fighters. People who join the fire service are clearly convinced of the value of putting out fires in chemical factories and rescuing children from the top floors of blazing apartment buildings. The greater vitality provided by SENS will translate into more children rescued and blazes doused. Fire-fighters will no longer be relegated to deskwork when they hit sixty. Their rejuvenated limbs, hearts, and brains will enable them to keep on dousing raging infernos until they become bored and retrain as mountaineer rescuers.

The question we must ask is how negligibly senescent fire-fighters will balance these increased potential losses and benefits. Though it's unlikely that every fire-fighter will make the same choice, I suspect that, were he to act as he predicts, de Grey's fire-fighter correspondent will be in the minority. SENS will make the vast majority of actual members of the fire service, and candidates for entry into it, significantly more risk-averse. There are clues about the future behavior of fire-fighters from present and past trends in soldiering.

When soldiers have been empowered to make choices about the conditions under which they fight it seems they've tended pretty consistently to place improvements in safety ahead of expanded opportunities for heroism. It may seem a strange word to use in this context, but wars have become progressively "safer" for the soldiers who fight them. Britain suffered close to one million military deaths during the four years of World War I. Admittedly, the number of British soldiers committed to Iraq was much smaller, but the military death toll for five years of fighting from March 20, 2003, until March 26, 2008, was 176. Multiplying that number by ten or one hundred would leave Iraq's death toll well short of the number for World War I, but it would dramatically exceed a modern liberal democracy's tolerance for military deaths. The boost in risk-aversion seems to have reduced the military effectiveness of the armies of modern liberal democracies. Tactics such as suicide attacks and roadside bombs, which have been every effective in Iraq, are unlikely to have made a significant impact on General Douglas Haig's British army in France during World War I. There'd have to be a pretty impressive density of roadside bombs to get close to the risk of death one would face from charging into no-man's land.

Could this preference of modern soldiers have been predicted? Perhaps not. Suppose you were to get into a time machine and go back to the era of Homer. You describe to him the various enhancements in military hardware that have occurred in the two and a half thousand years since his time. It's possible that Rambo would be more in keeping with Homer's prediction about the behavior of early twenty-first-century warriors. In Homer's stories the healthiest and fittest—Achilles and Hector, to give just two examples—were the most reckless. Yet few American or British soldiers serving in Iraq in the early twenty-first century seem to have taken Homer's heroes as role models.

What Makes SENS Different

In a discussion with de Grey, physicist David Deutsch sought to justify the heightened risk-aversion brought by SENS with the point that historical increases in life expectancy have tended to make us more risk averse.[17] As we've extended our lives, we've gradually become less tolerant of threats to them. A resident of medieval Europe had a life expectancy of between twenty and thirty years. Advances in nutrition, sanitation, and medical knowledge between then and now have added around fifty years to this figure, readjusting our perceptions of danger accordingly. It's difficult to think of activities routinely practiced by citizens of the modern rich world with levels of danger equivalent to joining a medieval duke's army or giving birth in the 1300s. Crop failures and visitations by the black plague were fixtures in the lives of medieval Europeans and would have helped them to place these risks into context.

SENS, however, is actually very different from the advances in nutrition, sanitation, and medical knowledge that have incrementally extended our lives. Recall how de Grey commenced the first of his talks to Google Corporation: "The first 1000-year-old is probably less than 20 years younger than the first 150-year-old."[18] We're comparing a twenty-year period in which life expectancies may improve by eight hundred and fifty years with the seven hundred years it took to add fifty years to our life expectancies. With the introduction of SENS, there will be abrupt and dramatic increases in our perception of the danger associated with many everyday activities. We won't reflect on some of the more dangerous of our familiar activities and decide that they're now a little bit too risky; rather, familiar activities will go quite suddenly from being easily safe enough to being far too dangerous.

At another point in his discussion with Deutsch, de Grey amends his earlier claim that driving would be outlawed. He replaces it with a prediction that we'll all be driving "incredibly safe cars with very sophisticated sensors that can avoid serious injury in the context of really massive human error."[19] I suspect that negligibly senescent people will in fact demand quite unprecedented advances in automobile safety. Air bags reduce your chances of dying in a car crash by approximately 8 percent. That's good enough to keep us senescing types on the road. But it's nowhere near enough for negligibly senescent people. They'll insist on titanium-plated, radar-equipped vehicles with top speeds of 10 kilometers per hour. Better still, they'll stay at home.

None of this should be taken as rationally compulsory. It's possible that negligibly senescent people will decide to treat the interior causes of death, such as the seven deadly things, differently from exterior causes of death such as driving, bush walking, and skiing. The one-thousand-year estimate of the life expectancies provided by SENS is, after all, based exclusively on the elimination of the seven deadly things. It does not assume that we've done away with any of the mortal threats posed by cars and tsunamis. Is there anything wrong with pronouncing yourself content with the mere one-thousand-year expected increase?

Though there's nothing wrong with saying this, I doubt that people susceptible to de Grey's rhetoric will be inclined to agree. It sounds suspiciously reminiscent of the "pro-aging trance" that encourages us to accept the risks of death that come from the seven deadly things. Those who take de Grey's therapies have quite deliberately set themselves on a course of reducing the probability of death in succeeding units of time. It's not hard to imagine de Grey insisting that trips to Egypt to see the pyramids and cycle-tours of Umbria can kill—the act of scheduling one of these holidays increases the likelihood that you will not survive the forthcoming year. There's nothing good or glamorous about death, even if it's accompanied by nice scenery. Negligibly senescent people have shown themselves unsusceptible to the pro-aging trance in respect of interior causes of death. They'll be spending much of their time self-administering anti-aging therapies. It's hard to imagine them cavalierly dismissing similar risks accompanying downhill skiing and helicopter flying. De Grey imagines negligibly senescent people becoming rock stars; but the caution that comes with negligible senescence seems antithetical to rock stardom as we currently understand it. There's unlikely to be any stage-diving into mosh pits, for example.

Perhaps there's an upside to this heightened perception of risk. Remember that Bernard Williams thinks that those who live past three hundred may find life incredibly boring. Boredom and the fear that I've argued may accompany millennial life spans could cancel each other out. Negligibly senescent people may tire of holidaying in parts of the world in which the combined threat of death from pilot error, natural disaster, or civil strife does not exceed a certain threshold. They may derive a visceral thrill from reckless acts such as being a passenger on a jetliner or visiting India. Their interest in these activities will be quite different from ours: We don't view driving as recklessly negligent, something we're prompted to do out of desperation at life's increasing tedium. Driving is an activity that we can do relatively safely.

SENS-able Thrills

This should not be taken to mean that there will be no pleasures safe enough for negligibly senescent people. In fact, there are some activities that are dangerous for us that negligibly senescent people may practice with relative impunity. Negligibly senescent people will distinguish between *sudden* and *gradual* causes of death. Sudden causes of death, which include jumbo jet crashes and falls while climbing Everest, are beyond remedy by SENS. Many gradual causes of death are.

I suspect that negligibly senescent people will spend a great deal of time on the Internet. The principal contemporary dangers from the Internet derive from the sedentary lifestyles of dedicated surfers. SENS will offer answers for these. The arteries of lifestyle Net-surfers will no longer clog with cholesterol. Their midriffs will no longer accumulate fat from daily consumption of pizza. There are likely to be other compensations.

In the 1973 Woody Allen movie, *Sleeper*, Miles Monroe is awoken after two hundred years of cryonic preservation to discover that many of the people of 2173 are smokers. The reason?—smoking, together with deep-fried foods, steak, cream pies, and hot fudge, has been shown to be healthy. Late twentieth-century audiences laughed at the absurdity of this suggestion. But there may be something in Allen's prediction. Smoking delivers nicotine, thought to enhance powers of concentration, a benefit currently outweighed by the habit's many detrimental effects. Though SENS has no answers to head-on crashes at cumulative speeds in excess of 200 kilometers per hour, it could provide a definitive fix for all of the various kinds of damage done by cigarettes. Smokers will feel particularly encouraged by WILT, which should reduce the habit's cancer risk to zero. Any activity that we currently recognize as dangerous but which does its damage slowly is likely to be safe for the negligibly senescent. They will have ample time to apply the rejuvenation therapies that emerge from SENS. In addition to smoking, I predict that Krispy Kreme donuts, chicken-fried steak, and chronic heavy drinking may experience renewed popularity. Negligibly senescent alcoholics won't own cars, so drunk-driving shouldn't be an issue. The brain atrophy and liver damage they suffer will be reparable by stem cell therapies. So there will be some fun activities to replace driving, sailing, bush walking, and wind surfing that they've had to renounce.

Smoking and drinking may be safe activities for those who already have access to the appropriate rejuvenation therapies. But those aspiring to millennial life spans should definitely defer these habits until after they have these therapies in hand. Though chain-smoking and binge-drinking may not shorten the life spans of the negligibly senescent, they have a proven propensity to make definite life spans shorter.

The Second-Hand Experiences of Negligibly Senescent People

So, some things that currently seem safe to us will come to seem too dangerous. Some things that currently seem too dangerous will become significantly safer. In this section I investigate a deeper consequence from this shift in the perception of danger. SENS is likely to alienate us from the things and people who currently give meaning to our lives.

Though I'm going to be making a prediction about the experience of negligible senescence, I certainly don't claim to be Nostradamus. I could be like the *Times* newspaper journalist who, in 1894, warned that, by 1950, the city of London would be drowning under horse manure. The journalist failed to predict the automobile, and I may be ignorant of future technological fixes for the problems I describe. In what follows I give reasons for thinking that such fixes may be difficult to find.

Currently, we get pleasure from direct, unmediated connections with parts of our environments. We swim in the sea, climb mountains, and ski down snowy slopes. Part of the pleasure we derive comes from the fact that they take us slightly out of our comfort zones—they seem a little bit dangerous. There are certainly places on our planet that are just too dangerous to experience directly—the insides of erupting volcanoes and the deepest ocean waters are two examples. But much of our world can be directly experienced by those prepared to put in the effort. The little bit of danger that makes snorkeling tropical reefs or flying a micro-light aircraft exciting for us is likely to translate into insane recklessness for negligibly senescent people. Negligibly senescent people won't be entirely excluded; they're likely to use technologies to gain access to these places. If direct contact with the sea or slopes is too dangerous, negligibly senescent people have a mediated access, perhaps via robot-mounted cameras. But we think differently about these kinds of indirect contact than we do about "being there." No one claims that seeing a Discovery Channel documentary filmed on Mount Everest substitutes for actually climbing it.

I predict that negligibly senescent people will retreat from the world. At a conference on transhumanism, de Grey let on that he would like to use the years he expects from SENS to go and see India. But modes of transport in India will have to become considerably safer before negligibly senescent Westerners deem them worth the risk. India's cultural and religious differences make it exciting to visit. They also make it dangerous, at least dangerous relative to the risk-tolerance of the negligibly senescent. Unless de Grey is willing to take such alarming risks, I suspect he may have to content himself with switching between "Exploring the Subcontinent" documentaries on the National Geographic Channel and reruns of the David Lean movie, *Passage to India*.

There's also a good chance that negligibly senescent people may have rather restricted contact with others. In our exploration of SENS-able therapies for intracellular junk we encountered the idea of the microbial infallibility hypothesis. It's the idea that if there's food around, some microbe will evolve the capacity to eat it. The microbial infallibility hypothesis is the source of de Grey's confidence that he can find microbes with a taste for the intracellular junk that causes atherosclerosis and Alzheimer's disease. But it also explains the ingenuity with which our microbial enemies find ways to circumvent antibiotics and other defenses we use to block their entry to our brains and bodies. At any given time, uncountable viruses, bacteria, and fungi are evolving new ways to consume us.

There's some reason to think that negligibly senescent people will be more vulnerable to infectious disease than we currently are. The reason for this has to do with two things that are features of the lives of senescing

people, but which will be largely absent from negligibly senescing lives—
death and sexual reproduction. It's possible that negligible senescence will
deprive us of one of our most powerful defenses against the microbes.

Sex perplexes evolutionary biologists. Sexual reproduction seems far
inferior in evolutionary terms to asexual reproduction. A successful asexual
reproductive act, making a clone of yourself, places the full complement
of your genetic material into the next generation. This seems twice as good
as a sexual reproductive act, which ferries only half of your genetic material into the next generation. So—given that sex is so reproductively inef-
ficient, how could it have evolved?

One of the prominent explanations has come to be called the *Red Queen
hypothesis*, named for the character in Lewis Carroll's *Through the Looking-
Glass* who has to keep on running just to be able to stay in the same place.[20]
Nature provides many examples of species that have to keep changing just
to stay in the same place. Consider the evolutionary arms race between
predator and prey. A predator species hits on a new technique to work
together to isolate a single prey animal from a herd. The prey species
responds by acquiring more acute eyesight and hence the ability to detect
predators at a greater distance. The predator species acquires a slightly
better camouflaged coat. The prey species counters . . . and so on. Both
predator and prey have to keep evolving just to stay in the same places
relative to each other.

According to the Red Queen hypothesis, organisms are in an arms race
against bacteria, viruses, and fungi that are seeking to parasitize them. The
parasites are constantly finding new ways to consume or get into host
organism cells, and organisms are constantly inventing new ways to keep
them out. The parasite–host arms race is actually markedly less fair than
the predator–prey example just described. Parasites are short-lived and
hence are able to evolve much more quickly than their longer-lived hosts—
they go through many generations for each one generation of their hosts.
Sex is the host's way of evening the odds. Bacteria and viruses attack our
cells by latching on to proteins on the cell surfaces. Sex produces novel
combinations of the genes responsible for putting proteins on cell surfaces.
This effectively changes the cellular locks. Better than just changing
them—it scrambles them. The parasite that is happily infesting mum or
dad is forced to evolve to crack your novel combination of cell surface
proteins. Death is actually a useful part of this response to parasites. It

retires those organisms whose cellular locks have been cracked, giving parasites no option but to evolve. Suppose our species were to do as the Raelian UFO cult insists, and trade in sex for cloning.[21] We'd find ourselves swamped by fast-evolving parasites ever more efficient at turning our bodies into food.

Negligible senescence is, from an evolutionary standpoint, a bit like asexual reproduction. It gives no opportunity to change the locks on cellular surfaces. In fixing the seven deadly things, SENS may increase our vulnerability to the quadrillions and quadrillions of parasites seeking to infest us.

Of course, we should not dismiss the resourcefulness of people who've already spent billions of dollars combating the seven deadly things. They won't be seduced by any lapse akin to the pro-aging trance and begin writing poetry about the loveliness of dying of typhoid. Unsurprisingly, de Grey is confident that negligibly senescent people will find effective responses to parasites. When I presented him with the problem, he proposed somatic gene therapy, gene therapy that targets the genetic material of body cells. De Grey's idea is that we could do for negligibly senescent people what sex does for senescing people. Regular bouts of gene therapy might change combinations of cell surface proteins at least as frequently as sex does for us.[22]

There are few problems that don't have logically possible scientific fixes. But we shouldn't fall for the same variety of technological optimism that describes logically possible technological fixes for global warming and asserts that we'll invent them, meaning that we can proceed merrily on our polluting way. We need to ask *how likely* we are to find these fixes—and *how soon* we'll find them. In the case of de Grey's response to parasites, we shouldn't make the mistake of understating the ingenuity of parasites. Human bodies are made up of a lot of stuff that they'd like to eat, and the microbial infallibility hypothesis supposes that they're very good at evolving ways to get at it. The genetic engineers tasked with implementing de Grey's plan are likely to run into difficulties. For example, there's good reason to think that changes to our cell surfaces that shield them against one parasite may increase vulnerability to others. In a 2008 paper published in the journal *Cell Host and Microbe*, Sunil Ahuja and coworkers report evidence that a gene that probably conferred resistance to a now extinct form of malaria actually increased the likelihood of its bearers contracting HIV.[23] Perhaps our microscopic invaders will happen upon ways to latch onto cell surface proteins that cannot be changed without adversely affecting the integrity of the cell. Or perhaps they'll evolve more direct ways to enter human cells. If we take the microbial infallibility hypothesis seriously we should not dismiss these possibilities.

It's possible that medical researchers will eventually outwit the microbes. They'll turn our cells into impregnable fortresses that will forever exclude viruses, bacteria, and fungi. One question concerns *when* this will happen. Remember that de Grey hopes that we'll achieve LEV in twenty-five to thirty years. The fact that we have only the vaguest idea about how somatic gene therapy will defeat the parasites should lead us to believe that it will arrive sometime after stem cell therapies or therapies to remove excess fat from our guts, for example. The arterial fat that leads to atherosclerosis is a breakdown in biological functioning that is difficult to fix. But the good news for our therapeutic efforts is that it's not a moving target. Fat has no capacity to evolve in response to therapies we invent. Microbes, in contrast, aren't a stationary, static enemy. They use natural selection to defeat our countermeasures.

There is one guaranteed protection against fungi, bacteria, and viruses. This is the strict avoidance of any sources of infection. The principal source of a microbe that is a threat to you is another human. Thus, negligibly senescent people may end up quarantining themselves from other humans, both senescing and negligibly senescing. They'll maintain barriers between themselves and others much like the condoms that reduce the likelihood of catching STDs. They'll meet only in Internet chat rooms. Since, if de Grey is to be believed, they'll have given up on reproduction, they'll be likely to view cybersex as a perfectly adequate substitute for any version of the activity that involves physical contact between human beings.

When I presented these concerns to de Grey, he was unfazed. According to him, "by definition something that hasn't yet killed the person you just met since they caught whatever they have won't kill you for a while either."[24] This sounds a little bit like an encouragement to have unprotected sex in the era of AIDS. Viruses do evolve to become less virulent. If HIV had quickly killed all those it infected it wouldn't have spread very far. The virus needs to balance its colonization of the infected person's body with the need to keep him or her turning up to the local discothèque. I suspect that negligibly senescent people will have plenty of opportunity to become infected with not particularly virulent but eventually lethal viruses. Their infected status will provide other negligibly senescent people with a further reason to avoid them.

None of this is to say that negligibly senescent beings cannot lead immensely worthwhile and rewarding lives. Negligibly senescent people are likely to find things to do that are both safe and more enjoyable than surfing the Net and chain-smoking. We may find the excitement of these activities difficult to grasp. But we might adjust. The species-relativism that I defend should make us skeptical about the value of this adjustment.

In this chapter I have challenged the more-of-the-same picture painted by de Grey and other advocates of radical life extension. Negligibly senescent lives are fundamentally different from those currently led by humans. In signing up for SENS we are, in effect, renouncing lives with the kinds of pleasures and pains we know well for fundamentally different kinds of lives with fundamentally different kinds of pleasures and pains. They might, like the very long lives of tortoises, contain greater numbers of pleasurable experiences than the lives of senescing humans. But those pleasures will not be ours. In chapter 9 I will have more to say about the species-relativism that vindicates rejecting greater quantities of unfamiliar pleasures in favor of lesser quantities of familiar ones.

Exploring the Social Consequences of LEV

Thus far we've been focusing on the experiential aspects of negligible senescence. Many of the things that we enjoy will be deemed unacceptably dangerous by people who legitimately expect to live for another thousand years. Negligibly senescent people will be downloading MPEGs of the Coliseum rather than venturing into a possibly structurally unsound two-thousand-year-old building. Their contacts with others may be mediated by the Internet. They'll prefer cybersex to sex. Conversely, hourly consumption of Guinness, cigarettes, and pizza may be life-shortening activities for us, but perfectly safe for negligibly senescent people. This makes the lives of negligibly senescent people stranger than we might think. I want now to explore a dangerous implication of the risk-aversion that comes for those who achieve longevity escape velocity.

The consequences of heightened risk-aversion extend beyond hobbies and holidays. As we've seen, many socially important activities are associated with a level of danger that negligibly senescent people should find unacceptable. Examples of socially valuable activities that are safe enough for us but may be judged foolishly reckless by the negligibly senescent are peacekeeping in war zones, rescuing drowning people, and putting out dangerous fires. There is likely to be some reconsideration of what kinds of activities should count as socially valuable. Negligibly senescent NASA employees will require considerably better justifications for climbing a mountain on Mars than "because it's there." Perhaps the idea of just wars will go out of fashion. Some commentators have observed that wealthy liberal democracies have great difficulty in prosecuting wars.[25] The young males who have traditionally made up the bulk of armies now expect more than half a century of mainly good quality life. They don't face imminent death from crop failure or the black plague and consequently find the levels of risk that soldiering entails unacceptable. If the trend toward increasing risk-aversion is at all indicative, the phrase "negligibly senescent soldier" may be a contradiction in terms.

It may be a good thing that negligibly senescent people are too fearful to go to war. Wars may not be quite as bad for us as they are for negligibly senescent people, but they're still bad. There is, however, one dangerous socially important activity of whose importance de Grey is emphatically convinced. This is SENS itself. Suppose de Grey's guess that we can reach LEV in twenty-five to thirty years turns out to be correct. Anyone who is fortunate enough to have access to the full range of rejuvenation therapies can rationally expect to live for a further thousand years. But it's an essential part of the concept of LEV that research should continue. The rejuvenation therapies available when LEV is reached will slow aging, but they won't stop it. If rejuvenation research stops or significantly slows then everyone's next millennium gets canceled.

The successes up to this point will, if de Grey is to be believed, have destroyed the pro-aging trance that is one of the principal obstacles to SENS in the early twenty-first century. The societies that we're imagining will commit massive resources to defeating aging. But the very success of SENS will place another obstacle in their way. This is the increasing risk-aversion of the beneficiaries of rejuvenation therapies.

New SENS drugs and procedures must be tested before their therapeutic efficacy can be confirmed. In this respect they're no different from new conventional medical drugs and procedures. The process of testing WILT, for example, is likely to be a multistage affair that will involve experimenting in various ways to genetically modify human cells, and procedures for infusing stem cells into the various parts of the body that need telomerase. It may begin with animal trials or very elaborate computer simulations. But new therapies cannot be pronounced safe and efficacious until they've been tested on humans. There are biological differences between apes and humans, and even very sophisticated computer models can fail to capture an aspect of human biology relevant to a therapy's efficacy. Candidates for negligible senescence want this testing process to be as expeditious as possible. Remember that it's the predicted pace of discovery of new rejuvenation therapies that determines whether at any given point in time we've made it to longevity escape velocity. Treatments for the remaining life-shortening damage must keep on coming, and the longer it takes to confirm the efficacy of new therapies with human trials the more likely we are to have to revise our claim to have accelerated to LEV. A delay in testing a new drug, perhaps motivated by concerns for the welfare of those on whom it's being tested, may mean that some aspirants to indefinite life spans will have to dramatically scale back their expectations. Rather than an additional millennium they may have to settle for a measly couple of decades.

Guinea Pigs for SENS?

I predict that there will be an obvious answer to the question of who will test rejuvenation therapies that arrive after longevity escape velocity.

My first point is that longevity escape velocity is very unlikely to be a single event. Different groups of people will reach LEV at different times. The judgment about whether any particular person has reached LEV must be made on the basis of the rejuvenation therapies that are available to him or her and those that are likely to become available in the future. In making this assessment, social and economic barriers are likely to be as significant as scientific ones. Recently there's been some debate about the extremely high price of cancer drugs.[26] For example, the breast cancer drug Herceptin (trastuzumab) costs around US$70,000 for a full course of treatment.[27] This means that qualifying for Herceptin depends as much on what you or your insurers can afford as it does on the particular biological characteristics of your cancer. If you don't qualify for Herceptin, then the drug's therapeutic powers should play no role in estimates about how long you're likely to survive. It matters not whether the reasons for your failure to qualify are biological—your cancer lacks the characteristics that make it suitable for Herceptin—or financial—you can't pay for it. If you're not taking the drug, it's not extending your life. The same points apply to rejuvenation therapies.

There may be a time when existing therapies and the expected rate of arrival of new therapies suffice to grant the wealthiest and best connected of us an additional thousand healthy, youthful years. People outside of this favored group may have some access to rejuvenation therapies, but only enough to extend lives by a few years. Perhaps the poorest people on the planet will still be pleading for therapies for HIV and malaria. They will have to wait a while for the rejuvenation therapies that have brought LEV to some to come down in price sufficiently to be available to them.

De Grey sets as a goal of SENS the "widespread availability" of rejuvenation therapies.[28] There's actually a big practical difference between the widespread availability that he hopes for and universal availability. We might avoid any problems arising out of graduated access by insisting that universal access be a precondition for anyone's getting them. This would guarantee that we all become millennial beings together. But I suspect that de Grey's coterie of wealthy supporters is unlikely to stand for this. Their contributions to SENS are made on the condition of the earliest possible access. They won't tolerate the delays entailed by demanding that the therapies be cheap enough to be made available to all upon first release.

As a result, this graduated arrival of LEV will create unprecedented differences in life expectancies. Those who have access to all the latest rejuvenation therapies will expect to live for an additional millennium. Many of those who don't will have life expectancies little changed from their socioeconomic counterparts today. I suspect that this vast difference in life expectancies will generate a moral asymmetry between therapeutic haves and have-nots.

World War I generals sent young men out of the trenches to do something that they deemed both important and far too dangerous to do themselves. Considerations about the respective potential losses of the young foot soldiers and the elderly generals make this seem back to front. The latter had considerably fewer years remaining than did the former and therefore had considerably less to lose from enemy machine guns.

Those who are separated by social or economic barriers from the full range of extant rejuvenation therapies will have less to lose than the potential recipients of the therapies. Risks that are intolerable to people with full access could be quite reasonable to them. In return for this service rendered to the negligibly senescent, they might expect better access to rejuvenation therapies. But many will have shortened life expectancies because of principled objections to SENS. I suspect that if de Grey has his way, people happy to continue senescing may not have much choice about whether they participate in experiments.

Will LEV Bring Medical Conscription?

De Grey wants SENS to advance as quickly as possible. To this end he argues against a variety of impediments blocking the progress of medical science.[29] One distinctively moral impediment arises in the attempt to protect human participants in medical research.

The twentieth century was a time of significant advances in medical knowledge. Some of those advances involved significant abuses of human dignity. These issues were brought into starkest relief by the activities of scientists tried at the end of World War II for "medical war crimes." Although "medical war crime" seems something of an oxymoron, there's no better term for the experiments conducted by Nazi researchers on concentration camp inmates.[30] These and other cases led to the formulation of strict protections for human participants in medical experiments. Researchers must respect the autonomy of human participants. This respect entails that potential participants be informed about the risks and benefits of any experiments and be fully empowered to give or refuse consent to participate in them.

I suspect that LEV could prompt a significant relaxation of the rules protecting human experimental subjects.[31] Those making and implementing the rules governing rejuvenation research after LEV are likely to be selected from the group that has the best access to anti-aging therapies. They know that unless the pace of research is maintained they will miss out on negligible senescence and will judge it imperative that therapies for the remaining damage arrive as quickly as possible. Expedited research may make the difference between their having an additional thousand healthy, youthful years, or succumbing to age-related damage that remains beyond treatment. Suppose a promising compound is rushed into human clinical trials too soon. This wouldn't be such a great tragedy. After all, in the worst case scenario, the people on whom it was tested may have lost a trifling thirty or forty years.

Consider how we might make the current situation of the testing of cancer therapies more analogous to the post-LEV testing of rejuvenation therapies. We would have to imagine that people suffering the earliest stages of a given cancer were given control over trials conducted on patients with more advanced disease. Those conducting the trials may make the case to the prospective participants that they have relatively little to lose from testing therapeutic compounds—they are, after all, dying of cancer. Those running the trials will correctly recognize themselves as having a great deal to gain if a miracle therapy is found.

I suspect that de Grey's notion of a war on aging, or WOA, may be instructive on the attitude that regulators of rejuvenation research may take. De Grey hopes that a declaration of war will inspire urgency. But the concept of war also provides pointers on how we are to go about ending aging. In war, lives are sacrificed to achieve objectives deemed very important. A general who refuses objectives that risk his or her soldiers' lives is unlikely to win any battles. We recognize the bravery of people who lay down their lives by calling them heroes and awarding posthumous medals. But these accolades and rewards are inadequate compensation for dying. Throughout most of human history, those who serve in armies have had considerably less choice about whether to fight than have those who lead them, and it's not surprising that coercion is sometimes required. Recruitment officers are asking young people to substantially raise their risk of death or serious injury. We need reassurance that there won't be some nasty form of medical conscription in which those who do not have access to the best rejuvenation therapies are expected to risk their lives in medical trials for those who currently have the best therapies but need even better ones to achieve their own personal millennia.

My principal focus in this chapter has been on aspects of negligible senescence that we might find unappealing. In the next chapter, I examine an argument that this bias in favor of the kinds of lives led by early twenty-first-century humans lacks rational foundation.

7 The Philosopher—Nick Bostrom on the Morality of Enhancement

So far we've focused on the technologies of radical enhancement. We turn now to the writings of transhumanist philosopher Nick Bostrom.

Bostrom's philosophical defense of radical enhancement is twofold. First, he claims that many arguments against enhancement rely on a fallacy. Identifying this fallacy has the effect of philosophically disarming many of enhancement's opponents, opening up the options of radical intellectual enhancement and life extension to those who wish to pursue them. This leaves unanswered the question of whether we should actually *want* to radically enhance ourselves. There are many options that should be available to the citizens of a liberal society but which few would want to exercise—covering your body in Elmo tattoos and attempting world records in baked bean consumption are two examples. Bostrom's second argument purports to close this gap between something's being permissible and its being desirable. Radical enhancement, on this view, turns out to be an unobvious but nevertheless direct implication of our shared human values.

I'll reserve discussion of the social consequences of radical enhancement for the chapter that follows this one. My chief critical focus in this chapter will be Bostrom's claim that human values are tacitly posthuman. I'll argue that the thorough investigation of our values that Bostrom is calling for is more likely to tell against radical enhancement than in favor of it.

Status Quo Bias and the Reversal Test

In an influential paper cowritten with Toby Ord, Bostrom diagnoses a mistake common to arguments against enhancement.[1] This is *status quo bias*, the error of viewing one option as better than another simply because it preserves the status quo. According to Bostrom and Ord, some of the seemingly most sophisticated bioconservative arguments boil down to the assertion that current human intellects and life spans are best because they're what we have now.

Bostrom and Ord's investigation of status quo bias draws on an extensive psychological literature. It opens with an experiment described by Thomas Gilovich and coworkers:[2]

The Mug Experiment.—Two groups of students were asked to fill out a short questionnaire. Immediately after completing the task, the students in one group were given decorated mugs as compensation, and the students in the other group were given large Swiss chocolate bars. All participants were then offered the choice to exchange the gift they had received for the other, by raising a card with the word "Trade" written on it. Approximately 90 percent of the participants retained the original reward.[3]

This pattern is a curious one to someone who starts with the idea that the decision of whether to trade or retain is motivated by judgments about the inherent values of mugs and chocolate bars. If either mug or chocolate bar was obviously of greater value than the other, then we would expect to see the participants to prefer it regardless of whether they started with it. Rerunning the experiment with a complete Wedgwood china set substituted for the chocolate bar ought to produce an obvious pattern, with those who started with the mug determined to trade and those who started with the china set resolved not to. This is not what the experimenters saw. But the pattern they witnessed also seems difficult to square with the assessment that the chocolate bars and mugs were judged to be of approximately equal value. If this were the case, then we'd expect to see roughly equal rates of trading and retaining, as the idiosyncrasies of the participants led them to prefer one reward to the other. One explanation that fits the mysterious pattern is a bias toward the status quo. The participants desired to keep the objects they started with simply because they started with them. Many of those unwilling to part with the mugs would have been chocolate hoarders had they started with Swiss chocolate bars.

Actually, Bostrom and Ord don't think that status quo bias is the only explanation for the systematic refusal to trade. They concede that the students' choices may reflect something known as "the endowment effect." It is a fact about human psychology that we form emotional attachments to things that we are familiar with. This is no trivial or superficial fact about human psychology. We don't automatically volunteer to trade our spouses for alternatives who may possess the attributes we claim to value to an objectively greater degree. It's possible that the brief period of ownership before trading became a possibility may have been long enough for the students to form some subtle attachment to their initial award, making the pattern detected by the Mug Experiment a rational one. The fact that mugs and chocolate bars have similar monetary values means that even a somewhat weak emotional attachment may be strong enough to explain the observed pattern.

There's one sense in which emotional attachments are highly relevant to policymaking, but there's another sense in which they should be diligently ignored. A politician considering a new law should clearly be concerned about the citizenry's emotional reaction to it—for example, a law that provokes widespread disgust is likely to be widely flouted regardless of its objective merits. But the politician should also understand that morally evaluating a potential law is a very different exercise from registering her own subjective likes or dislikes. For example, it would be wrong for a nation's treasurer to give as her only reason for rejecting a policy change that could bring significant economic benefits to everyone, that she had become emotionally attached to the old way of doing things.

Bostrom and Ord offer other examples that they think are not complicated by the endowment effect and therefore more closely parallel moral judgments. One concerns the preferences of Californian electricity consumers:

Electric Power Consumers.—California electric power consumers were asked about their preferences regarding trade-offs between service reliability and rates. The respondents fell into two groups, one with much more reliable service than the other. Each group was asked to state a preference among six combinations of reliability and rates, with one of the combinations designated as the status quo. A strong bias to the status quo was observed. Of those in the high-reliability group, 60.2 percent chose the status quo, whereas a mere 5.7 percent chose the low-reliability option that the other group had been experiencing, despite its lower rates. Similarly, of those in the low-reliability group, 58.3 percent chose their low-reliability status quo, and only 5.8 percent chose the high-reliability option.[4]

The endowment effect doesn't seem to explain the preference by participants in the experiment for a continuation of their original mode of electricity supply. Indeed, it seems likely that most of the consumers were unaware of the relative reliability or cheapness of their supply before their participation in the experiment. Bostrom and Ord propose that the best explanation for the pattern of responses in this experiment is an irrational bias toward the status quo. The decisive motivation in the Electric Power Customers case appears to be a brute desire for things to stay the same.

So what can we learn about the issue of enhancing central human capacities from decisions about mugs, chocolate bars, and electricity? The chief focus of Bostrom and Ord's paper is the enhancement of intelligence. They observe that many people combine a marked aversion to measures that would reduce either our intelligence or the intelligence of our children, with a reluctance to endorse measures that would enhance intelligence. This pattern could result from status quo bias. Bostrom and Ord offer a test that will help us to decide whether this is in fact the case.

Reversal Test: When a proposal to change a certain parameter is thought to have bad overall consequences, consider a change to the same parameter in the opposite direction. If this is also thought to have bad overall consequences, then the onus is on those who reach these conclusions to explain why our position cannot be improved through changes to this parameter. If they are unable to do so, then we have reason to suspect that they suffer from status quo bias.[5]

Most of those who oppose plans to enhance human intelligence are also against things that diminish intelligence—they'd oppose policies requiring the administration of hammer blows to children's heads or the prescription of lead tablets to breast-feeding mothers. We could demonstrate the rationality of this pattern of preferences if we could give a good reason for thinking that intelligence is currently at precisely the right level, or that the costs of modifying intelligence were exceedingly high or changes were just too risky. If we cannot produce such a reason we stand justly accused of status quo bias.

Bostrom and Ord consider some arguments that could explain why human intelligence could be at precisely the right point. The argument from evolutionary adaptation appeals to the conditions for which human intelligence evolved. Natural selection has provided human beings with adaptations that are a good fit for our environments, and it is possible that changes to intelligence in either direction might move it away from that evolutionary optimum, making us less likely to prosper. Bostrom and Ord doubt that this provides sufficient grounds for people in the early twenty-first century to reject the enhancement of intelligence. They make the point that the conditions in which our brains and bodies evolved are somewhat different from those in which humans currently find themselves. Greater intelligence might not have been adaptive in the hunter-gatherer environment of the Pleistocene, but it is more likely to be so, and to promote happiness, in an environment in which success depends on manipulating computers and swiftly appraising stock market options.

Another argument considered by Bostrom and Ord concerns transition costs, costs that arise in the move from one situation to another. If we boost intelligence we'll need to rewrite school curricula. The raising of intellectually enhanced children by intellectually normal parents may cause tensions. Bostrom and Ord present a reworking of the Reversal Test to show that transition costs should not lead us to reject cognitive enhancement.

The Double Reversal Test is introduced by way of a story involving the contamination of our water supply with a toxic chemical. We discover that drinking the water causes mild brain damage to those who drink it. The bad news is that we have no alternative sources of water. Fortunately, genetic engineers have produced a safe gene therapy that will increase our intelligence just enough to offset the damage caused by the poison. Bostrom and Ord think it's obvious that we should be permitted to use the therapy. They continue their story by imagining that our water supply is gradually cleansed of the chemical:

If we do nothing, we will become more intelligent, since our permanent cognitive enhancement will no longer be offset by continued poisoning. Ought we try to find some means of reducing our cognitive capacity to offset this change? Should we, for instance, deliberately pour poison into our water supply to preserve the brain damage or perhaps even undergo simple neurosurgery to keep our intelligence at the level of the status quo?[6]

Bostrom and Ord think it would be absurd to answer in the affirmative. Suppose our decision not to poison the water supply leads to transition costs. It seems wrong to say that the hassle of having to rewrite some school books and arrange counseling for parents struggling to relate to their more intelligent children justifies deliberately damaging their brains.

One might point to the risks of enhancing intelligence as part of an argument for the intellectual status quo. Bostrom and Ord concede that there may be unexpected costs as a consequence of enhancing intelligence. But they urge us not to ignore unexpected benefits. In fact, Bostrom and Ord go so far as to say that "uncertainty of the ultimate consequences of cognitive enhancement, far from being a sufficient ground for opposing them, is actually a strong consideration in their support."[7] They imagine a group of australopithecines who have somehow worked out how to boost their intelligence to the levels of *Homo sapiens*. When deciding whether they should use the new technique, they should be aware that their australopithecine brains would be incapable of grasping the "qualitative changes in our ways of experiencing, thinking, doing, and relating that our greater cognitive capacity have enabled, including literature, art, music, humor, poetry, and the rest of Mill's 'higher pleasures.'"[8] Though it would be impossible to describe the appeal of these experiences to australopithecines, this enlargement of their repertoire of valuable experiences should provide some objective reassurance. Humans contemplating radical intellectual enhancement can be similarly reassured. There seem, according to Bostrom and Ord, to be good inductive grounds for thinking that intellectual enhancement beyond human levels will expand our repertoire of valuable experiences.

I am skeptical of the general idea that the "uncertainty of the ultimate consequences" of an action is ground for support rather than for opposition. It would lead to conclusions both unwanted and unwarranted in respect of global climate change, for example. When dealing with living systems there's a general reason that ultimate consequences of which we are uncertain are more likely to be bad rather than good. A living system is an object of design comprising parts that generally work pretty well. For example, a human organism is composed of a heart that pumps blood efficiently, kidneys that do a good job purifying blood, and so on. Random alterations to living systems and other objects of design are much more likely to make them work worse than to make them work better. How likely is bashing a laptop that refuses to open Word to make it work better? Evolution by natural selection beats these poor odds by retaining the exceedingly few random mutations that make an organism biologically fitter and dispensing with the many that have the reverse effect. A defender of cognitive enhancement may complain that her chosen effects are not random. She's not randomly altering our cognitive powers but rather deliberately making us more intelligent. The bad news for her is that the human mind–brain is an exceedingly complex system and the increase in intelligence is but one among many effects produced by her intervention.

Many of the other effects are random, at least with respect to her understanding. Naming one good outcome—increased intelligence together with an enlarged repertoire of experiences—and pronouncing the sum of known and unknown consequences good is a bit like singling out a beneficial effect of climate change—increased wheat production in Siberia— and forming a optimistic opinion of climate change as a whole.

The Reversal Test enables Bostrom and Ord to examine the morality of extending our life spans. Most opponents of life extension would reject the suggestion that we should deliberately shorten our life expectancies. They therefore owe us an explanation for why it is that current human life expectancies are optimal, something Bostrom and Ord doubt that they can provide.

We can generalize the point of the status quo test. It calls into question the preference for remaining human. Those who reject enhancement think that we shouldn't become better than human. But they also think we shouldn't become worse than human. They should say why the human level of achievement marks precisely the right point on a continuum of abilities.

What's ingenious about Bostrom and Ord's argument is that rather than negotiating intuitions about the desirability or otherwise of enhancement, as moral philosophers are wont to do, it identifies a scientifically confirmed error in their opponents' thinking. Bostrom and Ord do leave one kind of response open. The rejection of enhancement could reflect the endowment effect—the propensity to form emotional attachments to things with which we are familiar. In the Mug Experiment an emotional attachment could have been sufficiently strong to direct a preference for one rather than another of two gifts of approximately equal monetary value. We've been human for considerably longer than the participants in the experiment had owned their mugs or chocolates, so we might expect our degree of attachment to be correspondingly stronger. The strength of our emotional bond with our humanity may make it rational to prefer this state of being to some objectively superior posthuman state.

In chapter 1, I described a species-relativist view of value, according to which some experiences properly valued by members of one species can lack such value for the members of another species. In chapter 2 I argued that, though it is not logically guaranteed to do so, radical enhancement is likely to export its recipients from the human species. Species-relativists should, therefore, be open to the idea that humans might have a rational preference for objectively inferior human experiences.

Species-relativism has implications for laws restricting enhancement. Such laws needn't reflect a brute preference on the part of policymakers that human levels of performance remain unchanged, an unwarranted bias on their part toward the status quo. Their opposition would instead be based on the recognition that a law permitting radical enhancement might threaten significant human values.

The points in the previous paragraphs do not tell directly against Bostrom and Ord's Reversal Test. Rather, they limit its application. What I've been calling moderate enhancement—increasing one's intelligence to the level of a genius like Einstein, for example—is compatible with our humanity. If this is so, then we cannot appeal to values available only to humans to justify its rejection. The reversal test may therefore expose many arguments against moderate enhancement as irrational.

Do We All Want to Become Posthuman?

Writing independently, Bostrom presents an argument that could be an effective rebuttal of species-relativism. According to Bostrom, radical enhancement may be something that we all desire, without being aware that we do. It's something that is implied by our human values.

We are all familiar with the phenomenon of people not really knowing what they want, and that there are certain kinds of situations, dysfunctional relationships and morale-destroying work conditions, that lead people to make mistakes about their true desires. Perhaps all of us really want to be radically enhanced. Or, more specifically, perhaps if we really understood what it was that we wanted, then we would realize that these things could not be attained unless we radically enhanced ourselves.

To reveal what it is that we really want, Bostrom offers not Freudian psychoanalysis, but a philosophical principle that should help us to avoid certain errors in working out what we really want. Bostrom explains that "[o]ur everyday intuitions about values are constrained by the narrowness of our experience and the limitations of our powers of imagination," continuing that "some of our ideals may well be located outside the space of modes of being that are accessible to us with our current biological constitution."[9] He enlists a particular theory of value to show how our values might be secretly posthuman. This is the dispositional theory championed by the philosopher David Lewis.[10]

The grounding intuition for the dispositional theory is the idea that our values are basically things we want. We want good health, and this is what places it among our values. The dispositional theory responds to the recognition that we are sometimes mistaken about what we want. Consider the drunk who says she wants to drive home. Although this desire may be very strongly felt, it doesn't seem to correspond with her values. The dispositional theory explains why. It states that "something is a value for you if and only if you would want it if you were perfectly acquainted with it and you were thinking and deliberating as clearly as possible about it."[11] Full imaginative acquaintance with the effects of alcohol on driving would make our prospective drunk driver aware of dangers that she currently overlooks, leading her to desire not to drive. It doesn't matter for the dispositional theory that she is so drunk that she is incapable of even approximating to these ideal conditions. Furthermore, it isn't the theory's job to make predictions about what will happen. Plenty of people who shouldn't drive nevertheless do so. It's a theory about human values and not about human psychology.

The dispositional theory helps us to accept some things with which we may be unfamiliar as values. But it also instructs us to reject some of the values that we currently credit ourselves with. For example, you may pronounce yourself a fan of Wagner's *Ring Cycle* after listening to the couple of minutes of "Ride of the Valkyries" featured in the movie *Apocalypse Now*. Yet if exposure to the full fifteen hours would cause you to withdraw your endorsement, then the *Cycle* does not belong among your musical values even if you tell people that it does. Wagner's anti-Semitism may lead other music lovers to a hastily formed low opinion of the *Ring Cycle*'s aesthetic properties. If proper exposure to it would lead them to reverse this verdict, then the *Cycle* is among their values even if they think it isn't.

Bostrom uses Lewis's theory to adjust our values in the light of perfect acquaintance and the most extensive possible deliberation to argue that there is no difference between human and posthuman values. We've seen how the dispositional theory corrects the blind spots of music lovers. It can also overcome human deficiencies in the understanding of our values. Bostrom says "Some values pertaining to certain forms of posthuman existence may . . . be values for us now, and they may be so in virtue of our current dispositions, and yet we may not be able to fully appreciate them with our current limited deliberative capacities and our lack of the receptive faculties required for full acquaintance with them."[12] Lewis's theory shows that many things that are beyond our comprehension nevertheless fall within the ambit of our current dispositions. Not even Garry Kasparov could grasp the basic principles of eight-dimensional chess. But presumably he would enjoy it were he to be fully acquainted with it. The same may be true for moderately gifted chess players. If we were to be properly acquainted with the hideously complex symphonies produced by posthuman composers we would find them beautiful rather than unintelligibly complex rackets. Posthuman symphonies are, therefore, among our musical values. It seems only right that we should seek to modify ourselves so as to better appreciate these things that we really want.

Bostrom's account yields the surprising result that bioconservatives may be disguised defenders of radical enhancement. If we listen to what Leon Kass, Francis Fukuyama, and Bill McKibben say, then they are among enhancement's staunchest opponents—they're quite explicit about not wanting to radically enhance their intellects or extend their life spans. But if Bostrom is correct in his deployment of Lewis's theory of value, they may, after all, have posthuman values. Were they to have radical intellectual enhancement imposed on them, they would enjoy posthuman pursuits. This could make them radical enhancers, albeit of the self-denying, self-hating variety.

Why Human Values Really Aren't Posthuman Values

But there is something a bit fishy about Bostrom's posthumanizing of our values. Notice the impact of Bostrom's rendition of Lewis's theory on some of the values we claim. I find Bach's B-minor Mass to be a beautiful piece of music. We can imagine that posthuman appreciators of music may find it trite and so not value it at all. Or perhaps they will value it, but only as an inoffensive wee ditty.

Bostrom counters that posthumans may continue to value simple things. He says that "the ability to appreciate what is more complex or subtle should not make it impossible to appreciate simpler things."[13] Bostrom gives some examples: "Somebody who has learned to appreciate

Schoenberg may still delight in simple folk songs, even bird songs. A fan of Cézanne may still enjoy watching a sunrise." But there seems a big difference between Bach's Mass being valued in this kind of way and the kind of value that music lovers currently place on the work. It's fine for posthumans to pronounce Bach merely pleasant and inoffensive, but it seems wrong to require us to echo them. This valuation, while positive, mistakes the significance of the Mass to human listeners.

There's actually a big difference between the rationally perfect self that Lewis imagines us deferring to and the radically enhanced beings that Bostrom wants us to become. We can be pretty confident that there will never be a rationally prefect being—at least, no such being will ever exist in the natural world. Such a being is logically incapable of error. Even the intergalactic super-intelligence that Kurzweil thinks the law of accelerating returns is taking us to is not logically protected from error. The difference between the rationally perfect being that is a construct of Lewis's theory of value and the radically enhanced beings that Bostrom thinks we want to become is an important one.

Lewis is trying to elucidate the values we currently have—not the values we might hypothetically acquire after frontal lobotomies, or decades of heroin addiction, or radical intellectual enhancement. He explicitly warns against construing the exercise of imagining possible candidates for valuing in a way that changes what we value. Consider the drunk who currently desires to drive home but would not if she were to be fully acquainted with drunk driving's possible consequences. The important thing about this imaginative exercise is that it includes facts about human capacities and limitations. It's actually irrelevant how well rationally perfect beings—or posthumans for that matter—manage to combine cars and alcohol. They probably drive faultlessly under any circumstances. Directing questions about the rationality of driving at either of these beings gives the wrong answer about drunk driving. It leads to the absurd answer that the drunk should drive—precisely the conclusion that the dispositional theory is designed to avoid.

The correct way to apply the dispositional theory is to imagine the drunk's ideal self making judgments about whether the drunk as she is currently should drive. While her fully informed, perfectly rational self suffers from none of the driving-related cognitive deficiencies of the drunk, she is charged with making a decision about whether an otherwise competent human being temporarily handicapped by alcohol should drive. This way of asking the question enables us to exclude drunk driving from her values. The aesthetic sensibilities of posthumans are equally irrelevant to human musical values. The classical music fan who happens never to have heard Bach's B-minor Mass would like it were he to hear it. This fact is known to Lewis's fully informed, ideally rational judge, who also knows that humans don't want to listen to posthuman symphonies any more than two-year-olds want to listen to Schoenberg.

Bostrom speculates about the worth of a "friendly superintelligence grounded in human values that might be able to advise us on ethical and other issues."[14] You should listen to the advice of this assistant when it is informing you about implications of your values of which you may be unaware, but not when it seeks to impose its own values on you. Accept its advice about the potential threat to your health and liberty from driving drunk. Reject its views about the aesthetic merits of Bach.

When I put the reasoning in the above paragraphs to Bostrom he urged that I distinguish "the value of a musical composition from the value of the experience of listening to it."[15] He continued that we can continue to be "very mistaken about how much value a musical composition has, even after we've listened to it (great works of art have often been unfairly neglected for long periods). It seems plausible in that case that there might be compositions that no current human would be qualified to assess even if they had great aesthetic value. And the human experience of listening to a posthuman composition may well have zero or negative value."[16]

There clearly is a big difference between the value we place on *experiences* of art and the value we place on the works of art, themselves. The experience of listening to a sublime performance of Beethoven's *Ode to Joy* might be very negative if you're dealing with a whopping hangover. In this case it's clear what would be the first change you'd have to make to shift your personal circumstances toward the ideal standard that determines your values. There are unjustly neglected works of art. Some of these are discovered long after their visionary writers, composers, sculptors, or painters have died. Doubtless there are some that are never discovered. They, like hideously complex posthuman works of art, are never appreciated by humans. What places them among our values is that they are *capable of being appreciated* by humans. There's a difference between the symphony that is never properly appreciated because no human judge ever bothers to listen to it properly and the posthuman symphony that humans could only ever recognize as good through the testimony of posthuman evaluators.

I suspect that full imaginative acquaintance with human values is likely to undermine the appeal of the radical enhancement sales pitch. In chapter 9 I'll explore some implications of their values of which many of radical enhancement's proponents will be unaware. One consequence of the species-relativism about valuable experiences that I defend is that radical enhancers may harbor a secret desire to retain their humanity.

So, Lewis's subjective theory of value is a false start for Bostrom. He might give up on subjectivism and argue instead that posthumanity is an *objectively* superior state. Humans may be prone to valorize being and remaining human. But perhaps these judgments are just wrong. By implication, perhaps species-relativism is just wrong. This version of objectivism about value seems difficult to sustain. For example, the experience of hearing a well-performed Schubert string quartet has high value relative to the aesthetic evaluative standards that humans tend to apply to music. It may have comparatively low value relative to posthuman evaluative standards. Who knows what value chimpanzee or dog or mouse listeners place on it. The complex symphonies that excite posthumans really are nothing more than rackets relative to the aesthetic standards applied by humans. All we can really say is that we don't like them but posthumans do.

A Cognitive Bias That May Support Radical Enhancement

We've seen that Bostrom and Ord accuse the opponents of enhancement of fallacious reasoning. I think it's possible that Bostrom's optimistic view of our posthuman futures may also rely on a fallacy.

Psychologists have explored a number of ways in which we err in our evaluation of future events. One kind of mistake is a consequence of the way in which we imagine the future. This error is known as *focalism*:[17] "People think about the focal event in a vacuum without reminding themselves that their lives will not occur in a vacuum but will be filled with many other events."[18] When we present future events to ourselves we tend to omit details. This would be fine if we didn't then tend to interpret omissions from these self-depictions as things that will not occur. The tendency to do so leads to mistaken "affective forecasts," mistaken predictions about how we'll feel when a specific possible future event happens. Timothy Wilson and colleagues offer the following illustration of focalism:

When people think about winning a million dollars, they probably imagine spacious mansions, round-the-world trips, and a cavalier attitude toward their children's college tuition. They might not anticipate the difficulty of maintaining relationships with envious friends, the hundreds of annoying phone calls from needy people seeking handouts, and the late-night worries about taxes and investments. The events that we imagine occurring are often quite different from the events that actually occur.[19]

In one experiment designed to investigate focalism, students from a US college were asked to imagine their team winning an upcoming football game and then make a prediction about how they would subsequently feel. They were first asked to make a prediction about how a loss would affect their emotional well-being. Interviews after the game revealed that students had overestimated both the ill effects on them of a loss and the good effects of a win. Their representations had systematically omitted both events that would cheer them in the event of a loss, and events that would bring them down in the event of a victory. They failed to predict that a loss would not prevent them from partying with their friends, and that a victory would not excuse them of having to study for midterm exams. Students tended to interpret these omissions from their imaginings as absences from their futures.

This propensity to omit detail becomes more pronounced the more distant the event. Wilson and colleagues say that "it seems likely that the focalism bias is worse when predicting the distant future. When people imagine how they will feel next week, they are likely to recognize that their lives will be full of other activities, such as their upcoming dentist appointment and soccer game. When imagining how they will feel next year, they probably think about the focal event in more of a vacuum, without considering the fact that their lives will be just as full then as they are now."[20] One study indicated that people are more likely to form "low-level construals" of events in the near future, construals that include information about other events, whereas they tend to opt for "high-level construals" of the distant future, construals that omit such information.[21] For example, when we imagine actions that may occur tomorrow we are more likely to represent information about how we will perform the action.

We are therefore more likely to think about obstacles that may arise or difficulties that we may encounter in performing the action. Descriptions of actions that we may perform in the more distant future tend to be pitched at a higher level, and are more likely to omit detail concerning how we will accomplish them. If you are imagining the seafood delicacy you will prepare tomorrow you are more likely to imagine how you might procure the ingredients, how early you will have to return from work to prepare it, and so on. You may alert yourself to obstacles such as the fact that your neighborhood fishmonger has been closed by the health department, and that the car has broken down, preventing you from traveling easily to a more distant store. The meal you are considering preparing for your spouse's birthday six months hence appears, in contrast, to miraculously appear on his or her plate.

I wonder if Bostrom's hugely optimistic evaluation of radical intellectual enhancement may be a consequence of focalism, with the added complication that our possible posthuman circumstances seem very distant, *pace* Kurzweil, from our current circumstances. Consider the following very vivid passages taken from one of his papers bearing the title "Why I Want to Be a Posthuman When I Grow Up." Bostrom begins by imagining the early stages in the ascent to posthumanity:

At the early steps of this process, you enjoy your enhanced capacities. You cherish your improved health: you feel stronger, more energetic, and more balanced. Your skin looks younger and is more elastic. A minor ailment in your knee is cured. You also discover a greater clarity of mind. You can concentrate on difficult material more easily and it begins making sense to you. You start seeing connections that eluded you before. You are astounded to realize how many beliefs you had been holding without ever really thinking about them or considering whether the evidence supports them. You can follow lines of thinking and intricate argumentation farther without losing your foothold. Your mind is able to recall facts, names, and concepts just when you need them. You are able to sprinkle your conversation with witty remarks and poignant anecdotes. Your friends remark on how much more fun you are to be around. Your experiences seem more vivid. When you listen to music you perceive layers of structure and a kind of musical logic to which you were previously oblivious; this gives you great joy. You continue to find the gossip magazines you used to read amusing, albeit in a different way than before; but you discover that you can get more out of reading Proust and *Nature*. You begin to treasure almost every moment of life; you go about your business with zest; and you feel a deeper warmth and affection for those you love, but you can still be upset and even angry on occasions where upset or anger is truly justified and constructive.[22]

Bostrom's imagination stretches to the even more distant posthuman future:

You have just celebrated your 170th birthday and you feel stronger than ever. Each day is a joy. You have invented entirely new art forms, which exploit the new kinds of cognitive capacities and sensibilities you have developed. You still listen to music—music that is to Mozart what Mozart is to bad Muzak. You are communicating with your contemporaries using a language that has grown out of English over the past century and that has a vocabulary and expressive power that enables you to share and discuss thoughts and feelings that unaugmented humans could not even think or experience. You play a certain new kind of game which combines VR-mediated artistic expression, dance, humor, interpersonal dynamics, and various novel faculties and the emergent phenomena they make possible, and which is more fun than anything you ever did during the first hundred years of your existence. When you are playing this game with your friends, you feel how every fibre of your body and mind is stretched to its limit in the most creative and imaginative way, and you are creating new realms of abstract and concrete beauty that humans could never (concretely) dream of. You are always ready to feel with those who suffer misfortunes, and to work hard to help them get back on their feet. You are also involved in a large voluntary organization that works to reduce suffering of animals in their natural environment in ways that permit ecologies to continue to function in traditional ways; this involves political efforts combined with advanced science and information processing services. Things are getting better, but already each day is fantastic.[23]

Is it possible that Bostrom is not only guilty of focalism bias himself, but wants to encourage it in his readers? The emotional coloration of his sales pitch for radical enhancement seems like that for a timeshare apartment in a tropical spot. In the latter case the salesperson wants you to focus exclusively on brilliantly sunny, carefree days lounging on the beach. He knows that we're likely to interpret absences from the imagined scenario as absences from the reality of the timeshare. Since you're not imagining tropical storms, congested airports, annoying bugs, and food-poisoning, he figures that you'll suppose that there won't be any. Likewise, Bostrom may be capitalizing on our tendency to read omissions from our imaginations of posthumanity as absences from posthuman realities.

Bostrom rhapsodizes about the wonderful experiences that radical cognitive enhancement would make possible. His willingness to part with wonderful experiences available only to humans in order to acquire them seems reminiscent of a home buyer prepared to purchase sight unseen. In chapter 9 I'll give a species-relativist reason for humans to remain human and to have children who are human. The reasons I provide will carry no weight with aliens or posthumans. This is because they do not purport to show that human experiences and achievements are objectively superior to alien or posthuman experiences and achievements. Before I present this argument, however, I investigate James Hughes's claims about the benign social consequences of radical enhancement.

8   The Sociologist—James Hughes and the Many Paths of Moral Enhancement

The *X-Men* movies depict the interactions between humans and a disparate collection of genetic mutants that have emerged, all of a sudden, from our species. The mutated DNA of this new breed of beings—the X-men—gives them a bewildering variety of somewhat implausible superhuman abilities, including mind-reading, the manipulation of metallic objects by thought alone, and miraculous self-healing. The humans are, in general, suspicious of these accidental posthumans and pass laws to keep them in check. Among the mutants, there are two schools of thought about humans. One view is represented by Magneto, played in the movies by Sir Ian McKellen. He doesn't believe in amicable relations between mutants and humans and forms his followers into a criminal gang to better exploit their diverse advantages over us. Xavier, a mind-reading mutant played in the movies by Patrick Stewart, believes in the peaceful coexistence of humans and mutants and strives to remove barriers to understanding between the enhanced X-men and the unenhanced humans.

James Hughes uses the *X-Men* movies to frame his discussion of the possible interactions between the beneficiaries of radical enhancement and those who remain unenhanced, either because they have not had access to GNR technologies, or because they have spurned them. Hughes sets himself the task of showing how we can head in the direction counseled by Xavier—that of harmonious relations between radically enhanced posthumans and unenhanced humans—rather than the Magneto path of conflict and exploitation.

The history of encounters between groups of differently empowered humans doesn't exactly inspire confidence in a Xavier future. There's a pretty consistent theme of dispossession, disease, enslavement, and murder that typically ends only because the oppressed acquire the technologies that are the source of their oppressors' power. The problem here is that posthuman advantages are not as straightforwardly shareable as iron-smelting and muskets. Various forms of injustice may therefore be permanent features of societies that include both unenhanced humans and radically enhanced posthumans, henceforth referred to as *human–posthuman societies*.

So, how does Hughes propose to ensure that human–posthuman relations are more harmonious than relations between human colonizers or conquerors and the peoples they colonize or conquer? The magic ingredient for Hughes is a concept of citizenship able to serve as a moral common ground between individuals who differ from each other in significant respects. He proposes to shift our understanding of what it means to be a citizen away from biologically parochial notions, such as humanity, toward the species-blind notion of *personhood*. The result will be a social arrangement Hughes calls *democratic transhumanism*. In democratically transhumanist societies, humans and posthumans will recognize each other as moral and political equals and act accordingly. Posthumans will make all of the GNR technologies available to humans who regret their earlier, or their ancestors', rejection of radical enhancement.

In this chapter, I argue that the precautionary approach recommends against the creation of any sort of human–posthuman society, something we can achieve by refusing to create posthumans. This conclusion depends on some speculation about the moral beliefs of posthumans that is, from the human perspective, less optimistic than Hughes's. Hughes offers a vision of *moral* enhancement. We will upgrade our current, flawed ethical codes to a morally improved democratic transhumanism. This chapter defends a skeptical view of Hughes's conception of moral enhancement. I argue that once posthumans come into existence, they may view humans as morally required to defer to them, to permit our interests to be sacrificed to promote theirs. Thus, the path of radical enhancement for some humans significantly threatens the interests of other humans. Though I don't pretend to be certain of this outcome, I am confident that a precautionary approach counsels against radical enhancement.

Avoiding Fukuyama's Scenario

Francis Fukuyama offers one of the more plausible rationales for the ill-treatment of unenhanced humans by enhanced posthumans. Fukuyama's warning is best understood in the context of his views about the connection between liberal democracy and human nature. Fukuyama became famous in the early 1990s for proclaiming the end of history.[1] This is not history viewed as the passage of time, which Fukuyama had no reason to doubt would continue, but rather history viewed as a series of often violent changes to the ways in which humans organize themselves into societies. As Fukuyama saw things, humans had progressed, or were progressing, through a series of despotic, oligarchic, feudal, and communist social arrangements, toward liberal democracy. The individual freedoms that characterize liberal democracy were, he thought, the perfect match for human nature. Humans who have found liberal democracy are like shoppers for trousers who have found perfectly fitting Levis. They have no need to keep looking.

Fukuyama thinks that the enhancement defended by advocates of radical enhancement will restart history by messing with one of liberal democracy's empirical preconditions.[2] Although some of us are more intelligent, healthier, or stronger than others, the differences are not so great that we do not recognize each other as potential contributors. Fukuyama quotes Thomas Jefferson's assertion that "the mass of mankind has not been born with saddles on their backs, nor a favored few booted and spurred, ready to ride them legitimately, by the grace of God."[3] This approximate empirical equality gives each of us the power to demand that others respect our liberty and moral worth. Fukuyama asks, "what will happen to political rights once we are able to, in effect, breed some people with saddles on their backs, and others with boots and spurs?"[4] We can imagine that supremely intelligent posthumans may see no value in liberal social arrangements that include those whose ancestors have rejected genetic enhancements and cybernetic upgrades.

George Annas, Lori Andrews, and Rosario Isasi go into greater detail on the violent consequences of creating or allowing the creation of radically enhanced beings: "The new species, or 'posthuman,' will likely view the old 'normal' humans as inferior, even savages, and fit for slavery or slaughter. The normals, on the other hand, may see the posthumans as a threat and if they can, may engage in a preemptive strike by killing the posthumans before they themselves are killed or enslaved by them."[5] Annas and colleagues propose that heritable enhancement should, therefore, be viewed as a "crime against humanity": "It is ultimately this predictable potential for genocide that makes species-altering experiments potential weapons of mass destruction, and makes the unaccountable genetic engineer a potential bioterrorist."

Enter James Hughes. Hughes shares Fukuyama's liberal political intuitions, but he thinks he can avoid the outcome Fukuyama describes. The trick is to replace our current biologically parochial concept of citizenship with one sufficiently inclusive to encompass both humans and posthumans. According to Hughes, a central tenet of liberal democracy is the moral and political equality of citizens. He thinks that this equality survives radical differences in intellectual and physical prowess. If humans and posthumans recognize each other as citizens then we may avoid futures in which the latter enslave the former and the former, anticipating this, attempt preemptive strikes against the latter. So where would this very inclusive concept of citizenship come from?

Our current attempts to forge inclusive moral understandings are clearly not up to the task. According to the Universal Declaration of Human Rights, adopted by the United Nations in 1948, all human beings are "born free and equal in dignity and rights." The declaration's timing is indicative of its overriding motivation. The horrors of World War II were fresh in the minds of its framers, and they hoped that a notion of universal human rights might help to avoid repetitions of humans shooting, bombing, or perpetrating genocide against other humans. But the declaration has a collateral effect that is less happy from the perspective of those who defend the moral standing of nonhumans. There's certainly nothing in the declaration that asserts that nonhumans are morally worthless. But those who choose to ill-treat them can offer the defense that their actions do not contravene the universal declaration. Posthumans are nonhumans. This means that the declaration affords them no protection. They may respond by inventing their own concepts of dignity and rights that offer no, or little, protection to us.

Hughes wants to escape this moral tit-for-tat by reconsidering what it means to be morally considerable. In place of a biologically parochial concept of citizenship centered on membership of *Homo sapiens*, Hughes proposes a concept that picks out attributes shared by humans and posthumans. This is the concept of personhood. Hughes adopts what has become the received philosophical analysis of persons as "self-aware beings with desires and plans for the future."[6] The important thing to notice is that no part of this definition mentions biological species. Some members of the human species qualify—but others do not. A fetus during its first trimester lacks self-awareness and must therefore be labeled a human non-person. Chimpanzees and gorillas are self-aware and entertain desires and plans for their futures and are, as a consequence, nonhuman persons. We saw in chapter 2 that radical enhancement may take our humanity from us. But it is unlikely to turn us into nonpersons. Posthumans would have to combine the radical enhancement of some central capacities with the radical diminishment of others to become nonpersons.[7]

Personhood is not just any characteristic that humans, great apes, and posthumans happen to share. It describes something of great moral significance. The mishaps and misfortunes of self-aware beings make a difference to them in a way that the mishaps and misfortunes of beings lacking self-awareness seem not to. Conveniently, personhood does not apply in degrees.[8] If you satisfy criteria for personhood, you don't become a person to a greater extent by becoming more self-aware or forming more complex plans. This makes the concept of personhood especially apt for establishing the equal moral status of apes, humans, and posthumans. We have higher IQs than gorillas. But a concept of citizenship founded on personhood will no sooner license an inference from the lower IQs of gorillas to their having a reduced moral status than it would justify granting a human with an IQ of 150 a moral status higher than that of a human with an IQ of 100. The IQ disparities generate differences in interests and levels of academic achievement, but they justify no differences in basic moral, legal, and political entitlements. Hughes's democratic transhumanism is essentially the suggestion that all persons be recognized as citizens and be granted equal protection under the law. The society he envisages is one in which chimpanzees, humans, and posthumans are acknowledged as moral and political equals. Recognition of this status would preclude bullying, enslavement, or any of the other abuses imagined by Fukuyama and others.

From this moral and political starting point quite a bit is supposed to follow. Hughes thinks that the citizens of democratically transhumanist societies will accept extensive responsibilities toward each other. They will support their fellow citizens in their attempts to "achieve their fullest capacities for reason, consciousness and self-determination." Hughes explains that "[w]hen our fellow citizens are less able, less healthy, less intelligent or less happy than they would otherwise be, it is our ethical and political responsibility to do what we can, while respecting whatever self-determination they are able to exercise."[9] This duty explains why we must commit resources to help "disabled citizens to achieve a fuller possession of their faculties for reason, autonomy and communication, so that they can control their own affairs." In the future, posthumans will understand that they have a similar moral responsibility to remedy human intellectual and physical deficiencies. They will help us to achieve a fuller possession of our faculties for reason, autonomy, and communication by uplifting us.

Uplifting is a notion prominent in the writings of the science fiction writer David Brin.[10] A biologically superior species uplifts an inferior one by enhancing its central capacities. Brin's novels depict a future in which humans uplift chimpanzees and bottlenose dolphins, transforming them into star-faring species. According to Hughes, the posthuman citizens of unequal societies will have an obligation to uplift humans should we request it. They may also have an obligation to realize Brin's scenario and uplift chimpanzees and dolphins. Humans differ from chimpanzees and dolphins in being able to request uplifting. Hughes asserts that uplifting could, nevertheless, benefit chimpanzees and dolphins. He makes the point that many of the conditions of zoo animals do not meet their needs. Intellectual enhancement would give them the capacity to better communicate these needs to us.[11] Hughes is aware of some of the potential complications. For example, he allows that "[w]e might find, for instance, that tweaking the intelligence of animals increases their sensitivity to pain or their neuroticism." He continues that "[i]f there were such a downside to the upgrades we should probably hold off, just as we would rethink sending children to schools if they became literate but miserable adults."[12] Hughes's example indicates how probable he finds these downsides of increased sensitivity to pain and neuroticism. Schooling does not, as a rule, lead to misery. And nor should uplifting.

The view that we are obliged to help our fellow citizens to "achieve their fullest capacities for reason, consciousness and self-determination" explains Hughes's confidence that the gaps in societies of the future may not be permanent. If posthuman lives seem so much better than our own human lives, we can easily reverse our earlier rejection of radical enhancement. The early adopters of radical enhancement will acknowledge a duty to help the rest of us to follow them.

A Precautionary Approach to Posthuman Morality

I find democratic transhumanism to be an attractive moral ideal. It is, however, important to note that it can only offer protection to the human members of human–posthuman societies if those societies realize its ideals. For example, the moral wrongness of murder and slavery did little to protect the Incas against the depredations of the Conquistadors. The issue I address in the following pages concerns the likelihood of democratic transhumanism becoming, or significantly informing, the *dominant moral codes* of human–posthuman societies.

A society's dominant moral code is the collection of moral ideas and principles that guides behavior in that society. It stands behind and justifies the society's justice system and the actions of its public officials. A dominant moral code may resist succinct summary. For example, there's no single ideal or principle that describes the dominant moral codes of modern Western liberal democracies. Rather, they comprise many ideas about the rights and duties of individuals and the importance of society as a whole, some of which can, on occasion, come into tension with each other.

It's inevitable that some of the very many moral ideas that form a society's dominant moral code will be erroneous. For example, they may make mistakes about the worth of certain of a society's members and what they owe to and are owed by others. Citizens are likely to have different opinions about which parts of their society's dominant moral code are mistaken and which are not. Some people have the misfortune to be members of societies whose dominant moral codes are built around ideas that are quite significantly mistaken—Nazi Germany and Apartheid South Africa are obvious examples.

This focus on democratic transhumanism is different from Hughes's. We will have only an indirect interest in the soundness of the moral reasoning that Hughes musters in support of the view. Democratic transhumanism will instead be viewed as a *prediction* about the dominant moral codes of human–posthuman societies. It can offer protection to humans who have missed out on or rejected radical enhancement only if it becomes, or significantly contributes to, the dominant moral codes of human–posthuman societies. Truths about the equal moral worth of human beings did not do enough to protect the interests of the black citizens of Apartheid

South Africa because they failed to find adequate expression in that society's dominant moral code.

We lack a principle to play the role Kurzweil's law of accelerating returns that would allow us to predict improvements in moral belief. There's obviously a considerable margin of error in any predictions that we humans might make about the moral views of beings with radically enhanced intellects. The following discussion should be viewed in the context of a precautionary approach to the creation of radically enhanced beings. In what follows I'll describe two distinctly human-unfriendly forms that posthuman morality might take. This book's precautionary approach means that I'll be challenging Hughes to show why the unfriendly posthuman moralities I describe are impossible or exceedingly unlikely.

Our current treatment of the great apes makes it clear that democratic transhumanism does not significantly contribute to the dominant moral codes of contemporary human societies. We destroy their habitats, confine them in zoos, and conscript them in medical experiments—all activities that contravene democratic transhumanism. So some degree of *moral* enhancement must occur to transform the moral codes that regulate human societies in the democratic transhumanist moral code that is to regulate human–posthuman societies.

Hughes is alert to potential problems for democratic transhumanism, and in a chapter of *Citizen Cyborg* entitled "Defending the Future" he recommends a variety of measures to ensure that posthumans give us our moral due. He follows Eliezer Yudkowsky in proposing that we design human friendliness into AI. He understands this as the idea that "values of tolerance and respect for human well-being are incorporated as core elements of the programming of machine minds."[13] Hughes urges that we design empathy, the capacity to imaginatively place oneself into the situations of others, feeling as they do, into our cognitively upgraded future selves and descendants. Empathetic beings try not to cause pain to others because they find doing so unpleasant. Hughes says, "Empathy is therefore intrinsic to the capacities that make us persons in the first place: self-awareness, self-concepts, desires and abstract reasoning."[14] This is the reason that psychopathic versions of posthumanity are unlikely to be attractive to those designing our descendants. Empathy is mostly an asset in social interactions, and not a handicap. Especially calculating psychopaths may thrive in the short term, but eventually they're found out.

The version of liberalism that Hughes endorses would permit the state to take steps to ensure that humans do not render themselves or their offspring devoid of empathy. He explains, "Just as we ask people to submit to driver's tests and ask potential employees to offer proof of citizenship, people could be asked for proof that they have fully functional capacities for empathy. Conversely, laws may be required to stop businesses and the military from selecting people with lower capacities for empathy, or providing technologies that facilitate the robotic pursuit of orders or profits."[15] Hughes recommends that we demand that "manufacturers of cognitive enhancement software . . . include empathy and moral decision-making supports as a feature just as we require warnings and child-proof caps on medicine and air bags in cars."[16]

Hughes makes the further suggestion that we develop cybernetic implants that run "moral reasoning support software" capable of informing us of what morality requires in any given situation: "We might then all be able to consistently reason with the clarity of philosophers and the selfless compassion of Gandhi or Martin Luther King."[17]

I think that there's some reason for optimism about moral reasoning of beings with radically enhanced intellects. Cognitive lapses are the cause of many human moral failings. We forget our resolution not to support factory farming when presented with the delicious-looking steak. We fail to appreciate that a resolution to help the world's needy encompasses the unprepossessing beggar on the street corner. Intellectually enhanced post-humans may be less prone to cognitive lapses like these. If they believe that humans are persons and that all persons are morally considerable, they may be less likely to overlook the implication that humans are morally considerable. Posthumans may be less susceptible to the motivational inertia that often prevents us from acting on our moral resolutions. It can take humans a while to renounce self-benefiting activities that we should recognize as immoral. Slaves existed long after the immorality of slavery was conclusively demonstrated. Perhaps there's reason to hope that posthumans will be faster than us at translating moral conclusions into moral actions.

I'm not persuaded, however, by Hughes's attempts to human-proof the posthuman future. It's relatively easy to say what would count as the enhancement of an integrated circuit. By adding transistors, we boost processing power and thereby enhance it. We have a good idea about what kinds of modifications to humans would enhance their cognitive powers. But we have a much less secure grasp on the idea of *moral* enhancement than we do on *cognitive* enhancement. In the following pages I address two possible patterns of moral enhancement that have implications for the human citizens of human–posthuman societies that differ markedly from Hughes's democratic transhumanism. If these forecasts are accurate, posthumans may cause suffering to humans not because they lack empathy, but because, like parents who present two-year-olds for painful vaccinations and cancer researchers experimenting on monkeys, they view that suffering as morally justified.

Two Alternative Moral Futures

In what follows, I assume that the moral notions prevalent in human societies do have some value for those trying to predict posthuman morality. Posthumans will emerge from us, so they're likely to start with our ideas about right and wrong. But these ideas are almost certain to change. This is partly because of the nature of moral truth. For example, our knowledge of chemistry tracks a mind-independent chemical reality. Posthumans may be aware of truths of chemistry that we're not clever enough to know. But their truths aren't different from ours. Transforming yourself into a posthuman doesn't make some former truths of chemistry false or some former chemical falsehoods true. Moral truths differ from truths of chemistry in being contingent, in some significant respects, on our psychological dispositions. The moral wrongness of murder, theft, and slavery must have something to do with a reliable tendency on the part of competent human judges to react to them in a certain way.[18] Changes to our psychological dispositions that occur as we become posthuman could, therefore, generate different moral truths.

My predictions of posthuman morality are based on extrapolations of two ideas prominent in human views about morality. These are the social contract view, which presents morality as an agreement between mutually disinterested individuals, and consequentialism, the idea that we should always try to bring about the best consequences. I will argue that posthuman developments of each of these families of ideas about morality may have negative consequences for humans. Posthuman morality may be very human-unfriendly indeed.

Our interest in social contract theory and consequentialism differs from that of moral philosophers who debate them. We won't be concerned about whether there's a statement of one of the two traditions capable of meeting all of the objections that human philosophers direct at it. Rather, we're interested in any information the theories might yield about the moral theories that will motivate radically intellectually enhanced posthumans. Posthuman morality may be a development of one of these traditions, but it is likely to differ significantly in detail from any version currently endorsed by human moral philosophers. How can we maximize the predictive value of contemporary statements of the social contract theory and consequentialism? The cognitive psychologist Gerd Gigerenzer has shown that would-be forecasters of intrinsically complex things such as stock market fluctuations or election results tend to err by basing their predictions on too much information.[19] More specifically, they tend to be influenced by information that is relevant to the past performance but which is likely to have minimal relevance for the future. Gigerenzer says that "in an uncertain world a complex strategy can fail exactly because it explains too much in hindsight. Only part of the information is valuable for the future, and the art of intuition is to focus on that part and ignore the rest."[20] I'm going to take Gigerenzer's advice and focus on the aspects of contemporary moral theories that are likely to have high predictive value. My forecasts will be based on generalities about moral traditions, ignoring much of the detail that interests contemporary moral and political philosophers. These very general characterizations will achieve enhanced predictive value through remaining indeterminate among a variety of alternative expressions of a given moral tradition.

Posthumans as Social Contract Theorists

According to the social contract view, morality is an agreement between selfishly motivated individuals. An individual complies with moral requirements because she recognizes that doing so is in her rational self-interest. Morality emerges from the recognition that we do better through accepting certain restrictions on our behavior than we would if we didn't. Moral rules that facilitate cooperation enable us to work together to achieve ends that none of us could achieve acting alone. We comply with a rule requiring us not to interfere with the liberty of others because we recognize that we benefit from others not intruding on our liberty. Among the prominent modern exponents of the social contract view are John Rawls and David Gauthier.[21]

It would go well beyond the scope of a book on radical enhancement to discuss all contemporary accounts, or indeed any of them, in detail. Remember that we are interested in social contract theory for any clues about the moral views of our radically enhanced descendants or future selves. Posthuman versions of social contract theory may differ significantly from any version defended by a human philosopher. I shall limit myself to indicating why humans should not view posthuman developments of social contract theory with any enthusiasm.

The concern for human moral forecasters is that posthuman social contract theory may endorse Fukuyama's scenario in which posthumans place saddles on the backs of humans. This outcome may be a consequence of a negotiation between posthumans and humans concerning the rules that will govern their treatment of each other. The moral equality that emerges from contemporary contractualism is premised on the idea that humans are roughly equal in terms of their abilities. We enter negotiations offering approximately as much as each other, and with roughly the same ability to deter ill treatment. If posthumans are much more capable than we are, then they may see less value than we do in a contract that limits what the powerful can do to the weak. The position of humans may be analogous to that of unskilled laborers who negotiate lower wages because employers view them as having less to offer than skilled counterparts. It may still be worth our while to enter into moral negotiations with posthumans. If they recognize our contributions as having some value, and that we have some capacity to interfere with their plans when completely ignored, then we may be able to negotiate a reduced level of moral recognition. We'll still, in effect, be offering up our backs for posthumans to place saddles on them, but at least we'll be able to impose some conditions on our service as beasts of burden.

Perhaps there is a way humans can emerge from a moral bargaining process with a status equivalent to that of posthumans. In John Rawls's version of the social contract theory we are supposed to imagine the contractors behind a "veil of ignorance":[22] They enter the bargaining situation unaware of facts about their current position in society, their talents, their ethnicity, their gender, their religious beliefs, or their views about the good life. This imagined ignorance ensures that the agreement they reach does not unduly favor the interests of some members of society over others. Perhaps we could add ignorance of our species to the list of attributes that contractors do not know. If you do not know whether you will enter society as a human or a radically enhanced posthuman, you might make rules that strictly limit the forms of exploitation that radical enhancement makes possible.

Although posthuman social contract theorists may endorse a theory like Rawls's, it is not clear that they would place one's degree of enhancement on the list of traits to be imaginatively ignored. The underlying truth of social contract theory is that moral rules exist to help us to achieve our own ends. There are beings with whom one enters into moral negotiations, and beings with whom one does not. For example, humans receive benefits from dogs—they rescue climbers, apprehend fugitives from justice, and keep us company. But we pay for these benefits by giving them food and shelter, not by granting them moral standing.[23] An offer of moral standing would exceed what is required to secure their cooperation. It may be true that were you to enter into negotiations about morality unaware of whether you were a human or a dog, you might agree to rules that strictly protected dogs. But Rawls certainly does not request that we imaginatively forget whether we are human or canine. Perhaps posthumans will be similarly reluctant to imaginatively overlook the many manifest differences between humans and posthumans. There are likely to be many differences between humans and posthumans that these future social contract theorists might invoke to make this exclusion seem principled.

Suppose, then, that posthumans grant humans moral status that corresponds with the value of their contributions. Things will start badly for the human members of a society run by posthumans, and they'll get worse. If Kurzweil is right, posthumans will continue to increase their powers while ours will remain substantially static. This means that the value of our contributions will decline relative to the value of theirs. Posthuman social contract theorists may reason that our moral status will decline as the value of our contributions declines. Eventually the gap may become so great that we may have nothing at all to offer. We will have no power to make moral demands of posthumans and posthuman social contract theorists may conclude, correctly, that humans have no moral value at all.

In conclusion, I think that we have reason to fear that a posthuman social contract will endorse some pretty shabby treatment of humans by posthumans. The significant fact about this shabby treatment is that it is licensed by the moral code to which posthumans may likely subscribe.

Posthumans as Consequentialists

A second possible dominant moral code for human–posthuman societies may offer hope for humans. Moral consequentialism is the idea that one should act so as to maximize global happiness and minimize global unhappiness. When facing a choice, one should determine which option leads to the best outcome when the costs of suffering are subtracted from the benefits of happiness. De Grey has defended SENS in an explicitly consequentialist way. He argues that money spent on radical life extension is likely to add greater numbers of healthy, youthful years to the global tally than is any other way of spending it. These inferior alternatives include current big-ticket items such as famine relief, international peace-keeping, and research on HIV.[24]

One reason for optimism about the implications for humans of posthuman consequentialism comes from the work of the best-known contemporary consequentialist, the Australian philosopher Peter Singer.[25] Singer uses consequentialist arguments to defend the claim that we must dramatically improve our treatment of nonhuman animals. We farm animals and conduct experiments on them in ways that cause them vast amounts of suffering. The human happiness gained from these activities is either of lesser magnitude (the pleasure gained from a new cosmetic does not compensate for the suffering of the animals it was tested on), or could be achieved in other ways that cause less or no animal suffering (vegetarian meals nourish humans just as well, or better, than meat, and they taste nice). This is what makes the testing of cosmetics on animals and the consumption of meat consequentially immoral.

If consequentialism does a good job of defending the interests of cognitively less capable animals against the claims of humans, then perhaps there's hope that posthuman consequentialism might protect the interests of humans against the claims of radically intellectually enhanced posthumans.

I suspect, though, that posthuman consequentialism is likely to come with a significant catch for humans. Suppose there were an omniscient, omnipotent consequentialist God. Such a being would presumably have the option of instantaneously boosting everyone's welfare to the maximum extent. Posthuman consequentialists may be more powerful than us, but they are not in the same league as the Omni-God. Like human consequentialists, they must make choices. If they decide to make some happier, they, in effect, forgo opportunities to make others happier. On other occasions, we recognize that we can make some happier, but only by reducing the happiness of others. The good news is that consequentialism gives clear and specific advice on how to make these trade-offs. We should choose the course of action that produces the best outcome when the costs of suffering are subtracted from the benefits of happiness.

Humans may not be impressed by the way in which posthuman consequentialists will make these choices. This is because posthuman consequentialism is likely to make the welfare of posthumans a higher priority than the welfare of humans. Consider the guidance Singer offers on how to make principled judgments about the moral priority of one kind of sentient being over another.[26] He proposes that we can find a "neutral ground" or "impartial standpoint" from which to compare a human and a horse. Singer describes a thought experiment in which he loses his humanity and turns into a horse: "And suppose that when I am a horse, I really am a horse, with all and only the mental experiences of a horse, and when I am human being I have all and only the mental experiences of a human being."[27] He continues, "Now let us make the additional supposition that I can enter a third state in which I remember exactly what it was like to be a horse and exactly what it was like to be a human being." This third state is somewhat peculiar, but it doesn't seem logically precluded, and so should be available for Singer's thought experiment. He says of the state: "In some respects—the degree of self-awareness and rationality involved, for instance—it might be more like a human existence than an equine one, but it would not be a human existence in every respect. In this third state, then, I could compare horse-existence with human-existence." Singer surmises that he might, from this third state, be able to decide whether he preferred human-existence to horse-existence, or vice versa. He says, "I would then be deciding, in effect, between the value of the life of a horse (to the horse) and the value of the life of the human (to the human)." He then makes a prediction about the outcome of this impossible experiment: "In general it does seem that the more highly developed the conscious life of the being, the greater the degree of self-awareness and rationality and the broader the range of possible experiences, the more one would prefer that kind of life, if one were choosing between it and a being at a lower level of awareness."[28] That is, we should take suffering caused to humans more seriously than apparently similar levels of suffering caused to horses. Humans have more to lose. By analogous reasoning, a third state in which it was possible to remember exactly what it was like to be a human and exactly what it was like to be a posthuman would lead us to place greater value on the "more highly developed conscious life . . . the greater degree of self-awareness and rationality and the broader range of possible experiences" of the posthuman than on the comparatively experientially impoverished life of the human. Posthuman suffering is, to posthumans, a more serious matter than is human suffering to humans. It follows that, in the eyes of impartial moral judges, posthuman suffering should matter more than human suffering. The same reasoning applies to human and posthuman happiness.

Singer's thought experiment may seem a bit odd. But consequentialists shouldn't be surprised at the conclusion it leads to. Radical intellectual enhancements are likely to give their recipients a greater capacity for happiness. The radical expansion of posthuman minds is likely to mean that they will entertain greater numbers of preferences about how their lives turn out. When we must choose, we're likely to most effectively maximize global happiness by focusing on posthumans.

Perhaps it's not so bad that morality has a bias toward posthumans. A human death may not be quite the same disaster as a posthuman death, but at least there is no doubt that it is a bad consequence, something that posthuman consequentialists must take account of. They are not permitted to disregard us in the way that posthuman social contract theorists within touching distance of the Singularity may choose to. Be this as it may, I suspect that humans should not be cheered by the suggestion that consequentialism will govern posthuman treatment of them.

Singer is right to condemn the vast majority of experiments on animals. Many cause great suffering while satisfying few human preferences that could not be satisfied in other ways. But Singer's consequentialism cannot support an unconditional prohibition of animal experimentation. In a recent televised discussion with Tipu Aziz, a prominent British researcher on Parkinson's disease, Singer conceded that it could be morally correct to conduct painful and lethal experiments on monkeys if the experiments bring us closer to a cure for the disease.[29] This apparent U-turn on the part of the author of *Animal Liberation*, the book that in 1975 provided the intellectual foundation for the animal welfare movement, was seized upon by Singer's opponents.[30] He later went to great pains to explain that his concession to the Parkinson's researcher was not a slip, but instead a simple implication of his consequentialism. In a letter to the *Sunday Times* newspaper he said: "Since I judge actions by their consequences, I have never said that no experiment on an animal can ever be justified. . . . If an experiment on a small number of animals can cure a disease that affects tens of thousands, it could be justifiable."[31]

Singer goes on to make clearer his complaint about Aziz's experiments. His worry is not so much about what Aziz is prepared to do—conduct lethal, painful experiments on rhesus monkeys to find a better treatment for a very nasty neurological disorder—but what he presumes Aziz would not do. Singer says, "In my book *Animal Liberation* I propose asking experimenters who use animals if they would be prepared to carry out their experiments on human beings at a similar mental level—say, those born with irreversible brain damage. I wonder if Professor Aziz would declare whether he considers such experiments justifiable. If he does not, perhaps he would explain why he thinks that benefits to a large number of human beings can outweigh harming animals, but cannot outweigh inflicting similar harm on humans."[32] Aziz stands accused of speciesism. This is, from Singer's perspective, the morally illegitimate privileging of human interests over the similar interests of nonhumans. Singer explains that "a prejudice against taking the interests of beings seriously merely because they are not members of our species is no more defensible than similar prejudices based on race or sex."[33]

There's an obvious reason Aziz is unprepared to publicly make the concessions that Singer would like him to make, regardless of his actual views on the matter. Consistent consequentialists must attend to the consequences of their moral pronouncements. There are occasions on which true moral pronouncements have bad consequences. A Parkinson's disease researcher who proclaimed himself prepared to experiment lethally on any irreversibly brain damaged humans who strayed into his care risks having his laboratory shut down. The consequences of conclusively demonstrating to Peter Singer that you're morally consistent are unlikely to be as valuable as finding a better treatment for Parkinson's disease. It may be a good thing that philosophers alert us to similarities between animals and cognitively subpar humans, but it would be bad, in our current circumstances at least, if these views were publicly defended by neurologists or neonatal pediatricians, for example.

Singer's consent to Aziz's research is too timid for a thoroughgoing consequentialist. He allows that Aziz's experiments on rhesus monkeys "could be justifiable." This seems to imply that we have some choice on the matter. Consequentialists should say, if it is the case that Aziz's experiments lead to the morally best outcomes, that the work is morally required. Compare: According to Singer, vegetarianism is more than just an option, a diet we can adopt if we feel like it, but not if we're too partial to steak and bacon. A meat-free diet is morally obligatory for all of those to whom it is available. This is because it's the choice that leads to the best consequences. Morally speaking, we don't have a choice.

The idea that it might be consequentially good for Parkinson's researchers to conduct painful and lethal experiments on rhesus monkeys is not implausible. The history of medicine is filled with advances that were made possible by experiments on animals. To give just two examples, large numbers of dogs died on the way to perfecting open-heart surgery and discovering insulin.[34] The hearts and pancreases of dogs are similar enough for experiments on them to be of value in the treatment of human cardiac abnormalities and diabetes. It wouldn't be surprising that research on neurological conditions might require subjects whose brains are more similar to our own than are canine brains—those of rhesus monkeys, for example. Mentally disabled humans may be even more useful research subjects. But a Parkinson's disease researcher's decision to experiment on humans is more likely to lead to a jail cell than it is to a medical breakthrough.

Of course, it's one thing to say that painful and lethal experiments on dogs have helped medical researchers. If we are consistent consequentialists we have to say that the same therapeutic results could not be achieved by measures that caused less suffering. For example, Singer doesn't deny that meat nourishes humans. Nor does he deny that it's important that humans nourish themselves. What he does claim is that a vegetarian diet nourishes in a way that does not cause large quantities of suffering to nonhumans. Herein lies the consequential superiority of vegetarianism.

Before consequentialists should countenance the killing of the rhesus monkeys, they would want assurance that the same discoveries could not be made in ways that caused significantly less suffering. For example, if there's a good chance that investigations of the corpses of Parkinson's patients would be just as effective at producing therapies, then it would be immoral to kill the monkeys. But the fact that a life-saving therapy might be found by other means isn't decisive. Consequentialists place a premium on efficiency. Suppose that the investigation of human corpses would lead to the same therapy as experiments on monkeys, but only after a considerable delay. Consistent consequentialists must take account of the suffering of Parkinson's patients between the time the discovery is produced by the corpse method and the time it would have been produced by the monkey method. Suppose that experiments on rhesus monkeys will yield a valuable Parkinson's therapy ten years before the corpse method. If the amount of suffering experimenters cause to rhesus monkeys is less than ten years of preventable Parkinson's suffering, then the monkey method may be morally mandatory.

The question we must ask is what purposes would posthuman consequentialists rate sufficiently important to morally justify the sacrifice of humans. Perhaps none. The radically enhanced intellects of posthumans may make them smart enough to avoid these morally tragic trade-offs altogether. They would presumably have no need for the morally wasteful varieties of medical experiment that human scientists perform. But it's certainly not impossible that there will be posthuman purposes that both require the sacrifice of human lives and lead to consequences sufficiently good to justify it. Increases in intelligence may help us do what we want less wastefully. But they also help us to find uses for previously unusable things. The current environmental crisis is partially a consequence of humans finding uses for parts of the environment unusable by our cognitively less sophisticated ancestors. We shouldn't presume too much insight into the designs of beings with radically enhanced intellects. But Kurzweil has been kind enough to suggest one. His theories about the Singularity give us some reason to believe that posthuman flourishing will require the biological deaths of humans. Kurzweil thinks that the law of accelerating returns will lead to the colonization of the universe by our minds. Every bit of matter and energy will become the substrate of, and fuel for, posthuman thought, and this includes the matter and energy that constitute human brains and bodies. These posthuman super-intellects may offer us compensation in the form of uploading. In chapter 4 I suggested that from their perspective uploading is likely to be preferable to biological survival, even if, from our human perspective, it is not. But they, not we, will be making the decisions. Should this super-intelligence be consequentialist, the relevant calculations are likely to point in the direction of enforced uploading.

Perhaps the above reasoning overlooks a crucial difference between the rhesus monkey subjects of human experiments and the humans whose bodies will be absorbed by posthuman super-intelligences. It's not surprising that rhesus monkeys, dogs, and seals come out on the losing end when their interests conflict with humans who happen also to be persons. Hughes's democratic transhumanism covers the interactions of different kinds of person, salient among them human persons and posthuman persons. Some moral theories do explicitly prohibit moral trade-offs. But consequentialists cannot unconditionally oppose them. Consider Singer's thought experiment: It is likely to place the "more highly developed" conscious lives, greater "self-awareness and rationality," and "broader range of possible experiences" of posthuman persons ahead of the "lower level of awareness" of human persons.

I'll conclude my discussion of consequentialism with an observation about the view's practical consequences. One of the most challenging aspects of consequentialism is the idea that it can be required to sacrifice the interests of some morally considerable beings to promote the interests of other morally considerable beings. I wonder how many of the philosophical advocates of consequentialism would be prepared to accept the practical implications of their theory when they get too close to home. For example, I imagine that it would take more than a rationally persuasive argument for moral consequentialism to convince an AIDS researcher to deliberately infect her children with HIV on the grounds that doing so will provide information about the virus's transmission that is likely to save hundreds of lives.

The beings that experience the harsh, unyielding side of consequentialism tend to be those who cannot speak for themselves. Medical researchers who inflict great pain on animals may be secure in the knowledge that the suffering they cause will prevent an even greater quantity of suffering. Human subjects are better models for human diseases than are the primates that currently fill medical research laboratories. Experiments on us may therefore actually have better consequences than experiments on non-human animals. They, not we, become research subjects, because they lack the opportunities for resistance that are available to us humans. Non-human primates may complain about painful procedures, but they are incapable of formulating these complaints in language that we recognize as moral.

Things may change when posthuman researchers rather than human ones are evaluating the morality of conducting lethal experiments on humans. Humans are certainly capable of making moral complaints. But their pleas may lack much of the moral terminology that posthumans use to conduct their moral disputes. They are unlikely to be supported by arguments that have the intellectual rigor and sophistication of posthuman arguments. Consequentially motivated posthuman researchers may take the same attitude to our complaints that human researchers currently take to the screeches of the rhesus monkeys they are experimenting on. They may proceed in the hope that the evident suffering they cause now will prevent more suffering later.

Again, we shouldn't be overly confident about the exact version of consequentialism that would be endorsed by radically intellectually enhanced beings. But if we are forced to make predictions about how posthuman consequentialists will treat humans, we should say that things don't look promising.

Does This Oversimplify Human–Posthuman Societies?

We can recognize Hughes's view of moral enhancement as occupying the most human-friendly extreme of views about the possible dominant moral codes of human–posthuman societies. Democratic transhumanism could feature strongly in the dominant moral codes of human–posthuman societies. But it might not. Other extrapolations of the dominant moral codes of human–posthuman societies are distinctly less friendly to humans.

When I presented this precautionary reasoning to James Hughes, he accused me of operating with too simplistic a notion of human-posthuman societies. If posthumans are clearly distinguishable in terms of their behavior and appearance from humans, then, he grants, there could be a "them and us" mentality. But Hughes thinks that the reality will be considerably messier than that, and that this messiness will be good for humans. He makes the point that "there will be a range of variation in degree of enhancement and a great diversity of enhancements," and continues, "That makes a scenario of cross-cutting identities and alliances much more likely than a radical discontinuity and speciation of post-humans who share nothing in common with humans."[35] Such cross-cutting identities and alliances are a feature of contemporary liberal democracies. Suppose Bob is a white, gay, Presbyterian, bridge-playing, global-warming-skeptic male. He has a multifaceted identity that permits and encourages numerous alliances with others in his community. This pattern explains the success of liberal societies; it prevents them from disintegrating into warring factions. Bob knows that his enemies on the climate issue may be his allies on matters of faith or sexuality, or his team-mates in high-level bridge tournaments. There's no reason that such alliances shouldn't ensure the success of human–posthuman societies. Gay posthumans will view themselves as more similar to straight posthumans in one significant respect, but more similar to gay humans in another, equally significant respect. And so on.

However, others present visions of human–posthuman societies that are altogether more schismatic than Hughes's. Lee Silver's forecast of a world populated by unenhanced human Naturals and multiple classes and sub-classes of enhanced GenRich is one such vision. He imagines diminishing communication and interaction between Naturals and GenRich as enhancement technologies become more powerful. According to Silver, the quickening pace of enhancement will move society toward "the final point of complete polarization."[36]

Although neither Hughes's nor Silver's visions of human–posthuman societies can be ruled out *a priori*, Silver's does seem the more likely to me. A posthuman with a radically enhanced intellect will be considerably more intelligent than a human being. Such a gulf in cognitive ability is likely to be a significant barrier to mutual understanding. I doubt that many gay men feel themselves to have much in common with a male macaque who either mounts or permits itself to be mounted by other male macaques. There's a good chance that posthumans whose sexual practices are the nearest equivalent to gay may experience a similar indifference toward homosexual humans.

There's a further point about how much these natural affections are supposed to count for. The members of a society may resemble each other in many different ways. A society's dominant moral code identifies certain of those respects as more important than others. You may share a star sign and hair color with your pet mouse, and these similarities may, in part, explain why you like the mouse so much. But you shouldn't make the mistake of viewing such similarities as placing you in the same moral category as it. You're a person and it's not.

Dominant moral codes are not defeated by natural continuities between members of one category and members of another. According to a view represented in the dominant moral codes of many modern societies, the act of killing an adult human being is very different from the act of killing an early human embryo. We recognize adult humans as morally different from early human embryos in spite of the fact that developmental continuities between the early embryo and the adult human make it difficult to precisely place moral dividing-lines. Intellectually enhanced posthuman philosophers may take a similar attitude toward the cognitive continuities that link them to human members of human–posthuman societies. Furthermore, their superior intellects may permit them to do a better job of superimposing moral distinctions on natural continuities than our moral philosophers do.

On the Moral Acceptability of Mere Existential Prejudice

There's a good chance that the dominant moral code of human–posthuman societies will be human-unfriendly. There are two measures we can take to ensure that humans are spared the misery of living in such societies.

Option one is to make radical enhancement mandatory for all. Those who are beyond compulsion would be eliminated. The first posthumans can consult movies in the *Terminator* franchise for ideas on how to do this. In this scenario everyone either travels the same path to posthumanity or goes out of existence—the arrival of the first posthuman would soon be followed by the disappearance of the last human. Human–posthuman societies would be a temporary stage on the way to universal radical enhancement. Option two is to ban radical enhancement. This is the option that my human moral intuitions direct me to.

Suppose we do resolve to prevent the existence of posthumans. This would involve two measures. First, we'd need to prevent people from coming into existence as posthumans. Second, we'd need to prevent humans from turning themselves into posthumans.

Preventing someone's existence is a very different matter from ending someone's existence. Whenever fertile couples choose not to have sex or to use contraception, they are, in effect, preventing the existence of possible people. Any duty to award the gift of existence to all possible people has intolerable implications. Even if our every waking moment were to be consumed by reproductive acts and encouragements to others to engage in such acts, we still wouldn't come close to awarding existence to all of the people who might exist.

We can legitimately prevent humans from having posthuman offspring or transforming themselves into posthumans by pointing toward negative consequences for those who choose not to follow this path. Our reasoning is essentially the same as that used to prevent humans from transforming themselves into psychopaths. We aren't talking about a universal ban on enhancement. We'd merely be saying that society has a legitimate stake in the enhancement agendas of its members. A commitment to the freedom of speech is compatible with a ban on varieties of speech judged injurious to society. By analogous reasoning, a commitment to the freedom to enhance can coexist with a ban on varieties of enhancement judged injurious to others.

What seems problematic is appealing to membership in a group as a justification for preventing someone's existence. Hughes picks up on this in his choice of the label "human-racist" for opponents of radical enhancement. He interprets human-racism as the idea that "only human beings can be citizens."[37] One can consistently seek to prevent the existence of posthumans while being disposed to grant full citizenship to any who come into existence. So this proposal is not human-racist in Hughes's sense. But perhaps it bears the taint of racism. When I recommend that we prevent the existence of posthumans I'm identifying individuals by their membership in a group. That certainly sounds like a form of prejudice. It singles out specific individuals for special treatment—in this case, nonexistence—based on their membership in a group—in this case, the group of posthumans.

We need some distinctions to get a secure grip on prejudice against posthumans. What I'll call *actual prejudice* takes as its targets the members of groups that exist. The familiar varieties of racism, sexism, or homophobia are cases of actual prejudice. The aim of what I'll call *existential prejudice* is to prevent the future existence of members of a particular group. Practitioners of genocide combine actual and existential prejudice—hostility toward actually existing members of a group and the desire to prevent future members of that group from existing. I'll understand *mere existential prejudice* as a variety of existential prejudice directed against a group that does not currently exist. There are currently no posthumans, so when I say that we should prevent their existence, I'm advocating mere existential prejudice against them.

One of the main reasons actual prejudice is bad is that it causes suffering. Racism, sexism, and homophobia are forms of actual prejudice that cause and pretend to legitimize great suffering. This is not a feature of mere existential prejudice. If the group whose existence you oppose does not currently exist, then it has no members to suffer harm. Incitements of violence against the members of nonexistent groups—humans with purple-and-pink-polka-dot skins, humans with eight limbs, and so on—lead to considerably fewer acts of violence than do incitements of violence against the members of groups who actually exist.

Practically everyone is guilty of some form of mere existential prejudice. For example, Hughes argues that we shouldn't bring into existence individuals with diminished capacities for empathy.[38] There is thus, trivially, a group—humans genetically modified to be exceptionally unempathetic—to whom Hughes would deny existence. He and I differ over the focuses of our mere existential prejudice.

The distinction between mere existential prejudice and actual prejudice may not be as sharply drawn as I've been suggesting. Consider the debate about the morality of reproductive cloning. Many nations have sought to prevent the existence of human clones by enacting legal bans on reproductive cloning. They thereby practice a form of mere existential prejudice directed against human clones. The legal scholar Kerry Lynn Macintosh suggests that this mere existential prejudice turns human clones into "illegal beings,"[39] and she sees problems with this prejudice. Macintosh makes the point that it is almost inevitable that, at some point in the future, human clones will exist. She thinks hostility toward reproductive cloning is therefore very like to give rise to actual prejudice against clones.

A law banning reproductive cloning criminalizes the act that brings human clones into existence, not the clones themselves. This may seem to be too fine a philosophical distinction to be workable as public policy. Is it asking too much of people to distinguish mere existential prejudice from actual prejudice? It's possible that our views about the immorality of human cloning will contaminate our views about the moral status of human clones and that we'll come to believe that beings whose creation was immoral have a reduced moral status. But there are cases that illustrate a weak connection between mere existential prejudice and actual prejudice. Consider a case of existential prejudice that takes as its target a group members of which currently exist. Rape is and should be illegal. To use Macintosh's terminology, children who result from rape are, and should be, illegal beings. We strive to prevent them from existing by preventing the act—rape—that is necessary for their existence. Nevertheless, we currently combine abhorrence for the act of rape with the recognition that children who result from rape have a moral status equivalent to those whose beginnings are consensual. I'm not aware of any move on the part of those who dedicate themselves to preventing rape to turn children who result from rape into second-class citizens. Someone who identified herself as a child of rape is likely to get a quite different reception at a meeting of citizens campaigning for tougher laws against rape than would a self-proclaimed rapist. Clones may suffer prejudice, should they come into existence. But this prejudice is unlikely to result from criminalizing the act that brought them into existence. If it exists at all, it will have other causes.

The position I'm defending doesn't make the move common in defenses of actual prejudice of presenting posthumans as necessarily vicious or stupid. The arguments I have explored in this chapter do not denigrate the moral worth of posthumans. Quite the reverse, in fact. They address their possible moral superiority. I allow that, should posthumans come into existence, humans may have to concede that their enhanced experiential capacities make it right that we suffer to promote their welfare. There would certainly be no justification for ill-treatment.

It seems to me that vampires are suitable targets of mere existential prejudice in much the same way as posthumans. In their canonical fictional portrayals vampires survive only by consuming human blood. Sometimes their blood meals lead to the deaths of the human donors. Sometimes the meals leave the human donors alive but significantly debilitated.

Though the vampire mode of nutrition causes suffering, there's nothing immoral, from the vampire's perspective, in the desire for human blood. A repeated refrain in fictional portrayals is that they didn't choose to be this way. Vampires clearly differ from humans who decide that they prefer human blood to other foods that could nourish them equally well. There are almost certainly combinations of the GNR technologies that could bring vampires into existence. If humans were, for whatever reason, to decide to choose to bring vampires into existence, then we'd be under some obligation to provide for them. This said, vampires are fit targets of mere existential prejudice. We should take steps to prevent their existence. We shouldn't be unduly concerned that a proposal to do so sounds, to some ears, prejudiced, bigoted, or racist. We aren't characters in fictions in which vampires exist.

Posthumans do not yet exist and it is up to us whether they ever do. If posthumans and posthuman interests never come into existence then we have nothing to place ahead of us and our interests. If they do come into existence then those of us who have chosen to remain human, or have had no option but to, may find the significance of our interests seriously diminished. Posthuman wannabes place the rest of us in an invidious situation. They want to create circumstance in which our interests, and the interests of our human children, are morally subordinated to their own or to their posthuman descendants. I think that we are entitled to prevent them from doing this. Those who choose to radically enhance themselves or their offspring may defend this choice as an expression of freedom over the state of their own bodies and brains or over the brains and bodies of their offspring. I have shown that it cannot be regarded as a purely self-regarding choice. It has potentially disastrous consequences for the moral standing of those who do not make similar choices.

9 A Species-Relativist Conclusion about Radical Enhancement

I've been arguing that the reality of radically enhanced intellects and extended life spans does not match their initial appeal. We should fear that Kurzweil's plan to make us super-intelligent and quasi-immortal will kill us. De Grey's therapies may give us more years but they'll also turn us into fundamentally different kinds of beings with very different fears and fancies. Bostrom's arguments concerning the desirability of radical enhancement and the irrationality of rejecting it fall short. Finally, Hughes's reassurances about relations between posthumans and humans work better as moral philosophy directing how we should treat nonhuman persons than they do as predictions about how posthumans will treat us.

So where does this leave us? At this point I'm in danger of seeming like a candidate for higher office who vigorously criticizes his opponents' policies but has nothing better to offer. In this chapter, I present some good news about remaining "merely" human. Humanity isn't just what we get left with once we've said no to Kurzweil, de Grey, Bostrom, and Hughes. It's something worth celebrating.

In what follows I explore implications of the species-relativist view about value described in chapter 1. I examine three kinds of valuable experience that are both typical of *Homo sapiens* and under threat from radical enhancement. The common theme is one of alienation. Radical enhancement alienates us from experiences that give meaning to our lives. My species-relativist presentation of these experiences grants that they are unlikely to appeal to posthumans or rational aliens. Our humanity should leave us equally unmoved by their posthuman experiences.

I'll begin by arguing that humans who undergo radical cognitive enhancement are likely to suffer a form of self-alienation. Cognitive enhancement can disrupt the proper relationship between us and the experiences and commitments that make us who we are. A second form of alienation affects our relations with our children. Some claim a right to radically enhance offspring as part of procreative liberty. I'll argue that the decision to radically enhance your children is likely to alienate you from them. Finally, I'll consider the ways in which we relate to those who represent the limits of what humans can achieve. I'll argue that enhancement alienates us from elite athletes. A significant human value directs us to prefer the more modest achievements of unenhanced humans to the superior performances of radically enhanced posthumans.

The Problem of Alienation

The advocates of radical enhancement are clear about what we stand to gain—dramatic enlargements of our minds and extensions of our lives. Those who oppose this transformation must locate losses sufficient to cancel these gains.

Francis Fukuyama takes up the challenge. He proposes that there are costs associated with the desire to eliminate aspects of our lives that we find unpleasant. According to Fukuyama, although "No one can make a brief in favor of pain and suffering," but "the highest and most admirable human qualities . . . are often related to the way that we react to, confront, overcome, and frequently succumb to pain, suffering, and death."[1] There are unexpected connections between the capacities, tendencies, and experiences that make up a human being. Fukuyama says that "Our good characteristics are intimately connected to our bad ones: If we weren't violent and aggressive, we wouldn't be able to defend ourselves; if we didn't have feelings of exclusivity, we wouldn't be loyal to those close to us; if we never felt jealousy, we would also never feel love. Even our mortality plays a critical function in allowing our species as a whole to survive and adapt."[2] Bill McKibben personalizes Fukuyama's point.[3] He writes touchingly about a childhood friend, Kathy, who suffered from cystic fibrosis. Kathy died when she was not quite fifteen, but in that short time impressed McKibben as one of the "happiest and kindest" people he had ever known. McKibben agonizes over the prospect of germ-line genetic engineering that might fix the cystic fibrosis gene. On the one hand, he's aware of Kathy's intense suffering. But this awareness is counterbalanced by the fear that Kathy's happiness and kindness might actually be in some sense a consequence of her disease. McKibben says, "Soon you're heading toward the world where Kathy's lungs work fine, but where her goodness, her kindness, don't mean what they did."[4]

Some of the apparent implications of this argument are difficult to accept. In a review of Fukuyama's book *Our Posthuman Future*, philosopher Colin McGinn observes: "Surely medicine is dedicated to the reduction of suffering, and it would be absurd to limit this effort for fear that people will become more superficial." He then asks, "Are we to nix a cure for baldness because we think thinning hair is character-building? And what about serious genetic defects that may also call forth impressive human virtues?"[5] Ultimately, Fukuyama and McKibben endorse the use of genetic technologies to treat or prevent disease. I think that they are best interpreted as challenging not the idea that we should be using technology to treat or prevent disease, but an overly simplistic view of the human significance of disease and suffering. For example, we can allow that a cure for cystic fibrosis would be overwhelmingly worthwhile, while acknowledging that distinctive varieties of goodness would disappear with it. Compare: World War II was, overall, a massively bad event. But it did enable some uniquely noble and virtuous acts. When we speculate about how wonderful it would have been had some disaffected Nazi assassinated Hitler in 1933 we're also allowed to think it a pity that Oscar Schindler would have been an indifferent businessman rather than the heroic savior of thousands. We can think this at the same time as finding it perverse in the extreme to block an assassin's bullet on the grounds that, although it was obvious that Hitler would cause great suffering, this suffering would elicit responses of great nobility.

The distinctive varieties of goodness occasioned by serious disease may be considerable, but their value doesn't justify keeping it around. Fukuyama's and McKibben's insights do, however, gesture toward something that is sufficiently good to justify rejecting radical enhancement. Disease and death present the opportunity for shared experiences. To use Fukuyama's words, all of us "react to, confront, overcome, and frequently succumb to pain, suffering, and death." If you receive a diagnosis of a serious disease, your first impulse is often to put yourself into contact with people already living with it. They may be the source of valuable advice on how to confront the forthcoming challenges. But the sense of belonging that derives from the shared experience has value independent of any useful information. Posthumans permanently immunized against serious disease may provide accurate scientific information about your disease's clinical progression, but they're unlikely to offer insight into what it's like to suffer it.

In what follows I describe experiences that differ from those of serious disease in being sufficiently valuable to warrant preservation. I give them a species-relativist presentation—they are consequences of the psychological and emotional commonalities that place all humans into the single biological species *Homo sapiens*. The best way to preserve these valuable experiences is to reject radical enhancement.

Cognitive Enhancement and Self-Alienation

Cognitive enhancement and life extension seem a natural pairing. It seems obvious that posthumans will not only be much more intelligent than us, they will live longer. I'll argue that those who realize de Grey's vision of millennial life expectancies have good reason to reject radical intellectual enhancement.

De Grey has a suggestive metaphor for SENS. He compares the repair and maintenance of the human body with the repair and maintenance of cars. De Grey says:

> Whether it's a poor laborer keeping the ancient VW bug running because it's the only car that he'll ever be able to afford, or a wealthy collector maintaining an old MG for the sheer love of it, we all know that a car can be kept going more or less indefinitely with sufficient maintenance. We don't have to keep the cars off the road in climate-controlled garages, and we don't rely on the latest gasoline additive: we simply repair worn-out parts when they begin to fail.[6]

De Grey's point is that a careful owner can keep a car going indefinitely. He wants to do to our bodies essentially what others have done to VWs and MGs.

We can learn something by pursuing the vintage car metaphor. As the years go by, the owner of the formerly top-of-the-line MG is likely to find that her car does not match the performance of newer sports cars. Of course, if she's a real MG devotee she doesn't care about this. Her devotion to her car is not premised on its being the fastest, most fuel efficient, or handling the best. Maybe it was in 1967. But it clearly isn't today. She resolutely rejects offers to replace her car's obsolete engine with a more modern one, instead taking great pride in keeping the original engine going. As parts wear out she is careful to source replacements that come as close as possible to matching the originals. I think that those who use de Grey's therapies to replace their worn out body parts may be as reluctant to accept bodily and intellectual enhancements as the car owner is to accept automotive enhancements.

Let's imagine that SENS has been a huge success. Longevity escape velocity has been achieved. You have access to the best rejuvenation therapies, and therefore expect one thousand more years of life. Life extension isn't the only area in which scientists have been making progress: They've discovered a way to radically enhance intelligence. Should you add this therapy to the many therapies you take to combat the seven deadly things?

One worry that you might have concerns your identity. The procedure that enhances your intellect will change the structure of your brain, leading to uncertainty about whether the person who emerges from the procedure is you. Some philosophers of personal identity counsel caution.[7] They defend a psychological view of identity that connects an individual's continuing existence to a continuity of mental life. According to this psychological view, your continued existence is explained by the persistence of memories, beliefs, and desires. Anything that threatens these psychological continuities threatens you. Radical intellectual enhancement is just such a threat to these continuities and, so the argument goes, should be resisted.

This objection to radical intellectual enhancement is only as persuasive as the view about our identities that it assumes. Bostrom and Ord think it's false. They compare the prospect of radical intellectual enhancement with the intellectual changes that occur as children grow up.[8] There are certainly major psychological discontinuities between five-year-olds and the forty-year-olds they eventually become. Regardless of what the psychological continuity theory might say, we think not only that we survive these changes, but that they are good for us. It's good to grow up, Michael Jackson's valorization of childhood notwithstanding. Bostrom and Ord extract an implication from our views about the goodness of growing up. If it can be good for children to grow up, "it might be good for adults to continue to grow intellectually even if they eventually develop into rather different kinds of persons."[9] We not only survive intellectual enhancement but benefit from its further opportunities for intellectual growth.

I don't share Bostrom and Ord's enthusiasm for continued intellectual growth—if by "continued intellectual growth" they mean ongoing intellectual enhancement. Suppose that radical intellectual enhancement experienced as an adult does leave your identity intact. The person who emerges with the four-digit IQ is you. I think you should find this a truly frightening prospect.

One of the things that people fear about diseases such as Alzheimer's is the intellectual decline itself. It progressively robs its victims of their reasoning powers and destroys their memories. But there's something else they fear as much and perhaps more—this is the severing of connections with the people and things that they value. People with advanced Alzheimer's may no longer recognize their spouses. They may no longer understand the social and moral causes that were among their strongest commitments. They may no longer remember that they were once presidents of the United States of America. In Alzheimer's disease the severing of connections is closely linked with intellectual decline. I think that significant intellectual growth may have a similar effect. It, like significant intellectual diminishment, has the propensity to sever your bonds with the things that really matter to you.

To see how this might happen, consider the implications of intellectual growth for ten-year-old Johnny. Currently, Johnny has a passion for model airplanes and Playstation games, especially those that involve fighter jets. He is shyly romantically interested in his school friend Mandy. This obsession seems to be based on the fact that Mandy is the only girl he knows who shares his aeronautical passions. One consequence of his ongoing intellectual growth that is unlikely to be apparent to him is that he is likely to lose the passion for airplanes and the accompanying romantic interest in Mandy. There are cases of childhood sweethearts whose mutual interest survives into adulthood. But these are the exception rather than the rule, much more consequences of good luck than good management. As we mature, our interests and passions change in ways that are difficult to predict. With increases in intelligence comes an expanded range of possible interests. A passion for model airplanes is replaced by one for eighties music, which in turn gives way to a passion for the works of Chaucer. A friend who at the age of twelve has interests that match yours is not likely to at age forty-two. The friend who engages your proto-erotic feelings at age twelve may not seem sexually attractive to you at ages twenty-two, thirty-two, or forty-two.

We accept the change in interests and attachments as part of the process of growing up. But eventually we reach adulthood, a stage that is supposed to mark the end of the dramatic intellectual and physical transformations that characterize youth. It also permits the formation of mature interests, interests that may persist throughout our lives. Of course, these interests are certainly not entirely static. They do change. But the change in our interests and commitment that we experience as adults differs from the change that occurs in the passage to adulthood. We may, as adults, tire of some of the things that formerly interested us. But these former interests do not become completely meaningless to us in the way that many of the interests of the child are meaningless to the adult. Adults don't just find the bendy animal figurines, squishy balls, and plastic trucks that fascinate young children uninteresting. The intellectual enhancement that they have undergone has completely alienated them from these toys. Your knowledge that the yellow plastic airplane might make a good gift for your two-year-old nephew comes largely from observing how other two-year-olds react to it, and remembering how the nephew reacted to similar toys. It's not a gift you buy because you find it engaging. You're likely to find it difficult to even imagine the two-year-old's excitement. Those who do buy two-year-old nephews gifts that they'd like bought for themselves are unlikely to make many favorite uncles or aunties lists.

Bostrom wonders whether we should be worried about the effect of becoming posthuman on the things we care about. He makes the point that we do not generally have commitments that will be impossible to fulfill as a radically intellectually enhanced posthuman.[10] Indeed, many projects may become easier to pursue. But the question is not whether they will be easier or more difficult to fulfill. It is whether we will desire to fulfill them. Adults can do many of the things that children do. They can stack blocks into towers and name all of the animals in a picture of a barnyard. They're actually better at these things than young children. The intellectual enhancement that occurs as we mature certainly does not take away our ability to perform these acts; it takes away our desire to do so.

Consider an adult faced with the same degree of intellectual growth as that typically experienced by children. We can predict that his interests and attachments would be similarly subject to change. Pursuits that currently fascinate him will be replaced by new, more intellectually involving pursuits. He currently has a passion for nature and cares about the consequences of global warming. A ten-fold increase in intelligence may not lead him to change his mind about the wrongness of pollution. But the commitment to nature is unlikely to occupy the same place in his moral consciousness as it did prior to the radical enhancement. A dozen or so more cerebral concerns may rate ahead of it—perhaps the global environmental crisis will be displaced by concern about a "galactic dark matter crisis." It is likely that his romantic attachments will be similarly subject to change. Currently he finds his partner's remarks about the fashion faux pas of friends amusing. After the next quantum of intellectual growth these observations are likely to seem banal.

Herein lies the threat to people with indefinite life spans from ongoing intellectual growth: It presents the prospect of never having any mature interests and attachments. This is a significant loss. We invoke a person's long-standing mature interests and commitments in explaining what defines her, what makes her distinctive. People whose indefinite life spans are accompanied by ongoing intellectual growth may, in contrast, present as a mutually unconnected series of commitments and interests.

You might take steps to reduce the risk that significant intellectual enhancement will sever your connection to those you currently love. Suppose you negotiate a his-and-hers deal at the enhancement clinic. You would enter the clinic as intellectual equals and emerge from it with that equality preserved. You might reassure yourself that people in relationships often grow together. Their beliefs and interests change in response to those of their loved ones. Perhaps this phenomenon of mutual change and growth means that you and your partner will form compatible posthuman interests, especially if you synchronize your cognitive upgrades.

I'm not so optimistic about the prospects for indefinitely long love affairs for people who continually enhance their intellects throughout their indefinitely long lives. Though it's possible that romantic relationships may survive radical cognitive enhancement, this is likely to be the exception and not the rule. Radical intellectual enhancement is likely to dramatically expand the range of things that you can become interested in. As you undergo radical intellectual enhancement, whether it takes place over the course of a few hours with the grafting of neuroprostheses or over several years with ongoing gene therapy, there will be many influences on your psychological development. The bond with your loved one is but one of these. Perhaps the likelihood that two companions will have compatible interests post-enhancement is about the same as the likelihood that the childhood friend who shares your passion for toy locomotives will be interested in the same things as you when forty. The couple who enter the intellectual enhancement clinic hoping that their intellectual and physical compatibility and mutual love will survive the procedure is a bit like a couple who ask their respective employers to relocate them, and just hope that they get sent to the same city. Those who truly value their romantic relationships tend not to undertake random changes and hope for the best. They protect their relationships against potential threats. Radical intellectual enhancement, like random geographical relocation, is one such threat. If you love someone, you reject the suggestion that you should permanently move away from him or her to a different city even if the move constitutes an enhancement of your residential circumstances, and, by analogous reasoning, you reject the prospect of intellectual relocation that might take you away from him or her. The same kinds of consideration apply outside of our connections with other people. To the extent that you really care about Bach's music, you should be suspicious of a transformation that will make it seem trite.

You may be reassured by the fact that you will, as an enhanced being, acquire new interests; though you are likely to lose any interest in the woman or man you are currently deeply in love with, you will find plenty of posthuman potential partners. But this possibility is only reassuring if you don't really care about the people with whom you currently share your life. The expression "plenty more fish in the sea" is typically presented to people who have loved and lost, not to people currently in love. I suspect that there is only one commitment that has a really good chance of emerging intact after the enhancement procedure. That is the commitment to enhancement itself.

When I put the preceding argument to the transhumanist philosopher Mark Walker, he was unconvinced. He presented me with a dilemma. Walker imagined a Peter Pan Pill that permanently arrests a child's development at age ten and asked whether I would recommend it. If one's current values are as important as I seem to be suggesting, then surely the answer is yes. Taking the pill preserves the child's current values. Suppose

I choose the more plausible alternative and oppose the Peter Pan Pill. In effect I would be saying that the child should trade her current collection of values for a superior collection of adult values. Yet this line of response seems to leave me vulnerable to a decisive rebuttal. Radical enhancement can then be defended on the grounds that as it disconnects you from your current human values it connects you with a new collection of superior posthuman ones.

Radical enhancers could take steps to ensure that their lives contain mature interests by deliberately limiting the number of times they undergo radical enhancement. Effectively we'd be presented with a choice between two different patterns of lives that both have mature interests. In human lives as they currently are, mature interests emerge after a period of dramatic cognitive enhancement we call childhood. Consider the following possible pattern for a posthuman life. Posthumans are born and live the first years of their lives approximately as we do. After a period of considerable intellectual growth they emerge into a stage that we would call adulthood. But for them this stage does not represent full maturity. Rather, it's an intermediate state on the way to a radically enhanced posthuman adulthood reached after the application of a collection of GNR technologies. The mature interests of posthumans form only after this transformation. This manner of posthuman existence differs from an indefinite existence with ongoing applications of increasingly powerful cognitive enhancement technologies in that it reserves a place for mature interests. Both human and posthuman lives are characterized by the formation of commitments from which one later becomes alienated. The posthuman life differs in arriving at a grander destination, and it seems preferable for that reason.

I think the human pattern is more valuable than the posthuman pattern. The value I'm talking about here should be understood in a species-relativist way. Human beings prefer it. Consider our reasons for rejecting the Peter Pan Pill. Children may not reflect on human development in quite the way that adults can, but, quite early on, they become aware that childhood is a temporary stage on the way to adulthood. It's the kind of existence enjoyed by their parents, caregivers, and teachers. They may be aware that their interests will change—it's apparent that their parents don't share their obsessions with Thomas the Tank Engine and Dora the Explorer. The Peter Pan Pill would leave them struggling with Mickey Mouse puzzles while their friends were enrolling at university. The important point is that we do not take this kind of perspective of our adulthood. We don't view it as a stage we're just passing through whose commitments are temporary and likely to be renounced. Of course, the values of post-humans will incline them to take a different attitude. They'd be as distressed about the prospect of being marooned in "second childhood" as human children would be about never getting to have experiences like those of mom and dad.

The Disconnect between Procreative Freedom and Radical Enhancement

Prominent among human values are relationships with family members. So it's worth asking to what extent, if any, they might constrain enhancement. Perhaps parents wouldn't choose to trade their moderately achieving child for a distinct academically outstanding one. But it seems that when they have the opportunity of enhancing without affecting the identities of their children, many parents choose to do so. Parents seek to enhance their children's intellectual powers by purchasing Baby Einstein DVDs and investing in after-school mathematics and French-language lessons. These practices seem to be powerful precedents for the more dramatic enhancements enabled by GNR technologies.

The advocates of radical enhancement have not been slow to recognize the seemingly widespread parental interest in enhancing their kids. They are foremost among those arguing that parents should be free to use genetic technologies to improve their children's characteristics. In his seminal treatment of the notion of procreative liberty, John Robertson defines it as "the freedom to decide whether or not to have offspring and to control the use of one's reproductive capacity."[11] Robertson proposes that enhancement should be recognized as a legitimate extension of procreative liberty. The advocates of radical enhancement have enthusiastically followed Robertson's lead. Gregory Stock presents technologies that enable the selection or modification of our genetic constitutions as germinal *choice* technologies.[12] Ronald Bailey indicates his liberal leanings in his selection of the title *Liberation Biology* for his book defending transhumanism.[13] He is a libertarian who combines defenses of individual choice about enhancement with skepticism about any role for the state. As we've seen, James Hughes's fusion of radical enhancement with social democracy differs; he emphasizes individual freedom but wants to allow the state to correct inequalities in access and to discourage individuals from making bad choices.

Nevertheless, there's actually a significant gap between the straight-forward endorsement of radical enhancement urged by this book's central figures and a liberal approach to it.[14] Classical liberals do not present themselves as marketing any particular view of human excellence. Rather, they defend institutions that allow individuals to make their own choices about how to live. Liberal pluralism about the good life carries over to decisions about what to view as an enhancement. The many different views about which is the best life lead to equally many views about what modifications to children's DNA actually enhance them. Liberals ask only that our choices be consistent with our children's well-being.

Consider the following case. Dean Hamer has recently argued that alternative versions of the gene, VMAT2, differently affect a person's propensity to hold religious or spiritual beliefs.[15] According to Hamer, people with one version of the gene have higher levels of a trait psychologists call *self-transcendence*. Self-transcendent types tend to see themselves as part of a larger whole. They're more likely than people who score lower for the trait to see a deeper significance in the car park on the busy road that becomes available just when it's needed, or in the phone call from the friend that happens just when you're thinking about him. People who have certain kinds of religious or spiritual belief may view replacing the low self-transcendence version of VMAT2 for the high self-transcendence version as an enhancement. And, so long as the resulting child wasn't harmed, this modification could be permitted by liberals. But it's unlikely to be similarly viewed by advocates of radical enhancement.

Advocates of radical enhancement differ from liberals in having definite views about the kinds of procreative choices that prospective parents should be making—they should be taking the first steps toward posthumanity, choosing, if possible, to have children who are much smarter, healthier, and longer-lived than ordinary humans. While liberals would protect the choices of prospective parents with posthuman values, they also want to protect, and indeed endorse, the different choices of parents who lack such values.

It is not hard to think of choices that would excite radical enhancers at the same time as being widely rejected by human parents given permission to alter their children's genomes. There seems a big difference, for example, between genetically altering Johnny so that he is ten IQ points smarter than he would otherwise be, and making him smarter than his parents to the same extent that they are smarter than nonhuman primates. The prospect of being viewed by one's child as permanently in the "da-da" stage of development would be a pretty terrifying prospect to many parents. We value being connected with our children, even if we know that severing this connection may provide them an objectively superior start in life. Few poor-world parents petition to have their children raised by the singer, Madonna, even if they believe that she provides material and educational advantages that they cannot match. We reject the radical enhancement of our children for the same reason that we don't offer them to Brad and Angelina. We value experiences shared with our children. We want to understand and take pleasure in their achievements, and we want our achievements to be meaningful to them. Neuroprostheses and genetic upgrades that radically enhance our children's cognitive powers place these shared experiences under threat.

Sport and Shared Human Experience

Does opposition to radical enhancement justified by human values extend beyond the self and family? I think that it shapes the interest we take in the exceptional physical and intellectual achievements that are part of elite sport. The massive global audiences of the Olympic Games and soccer World Cups are testament to the extent of our interest in the doings of elite athletes.

What is the basis of this fascination? Julian Savulescu, Bennett Foddy, and Matthew Clayton endorse the enhancement of elite sporting performances on the grounds that "the athletic ideal of modern athletes . . . is superhuman performance, at any cost."[16] This is not a view shared by the World Anti-Doping Agency (WADA), which strives—some say in vain—to keep performance-enhancing drugs out of elite sport. WADA has deemed performance-enhancing drugs and modifications contrary to the spirit of sport.[17] Savulescu and colleagues protest that "Far from being against the spirit of sport, biological manipulation embodies the human spirit—the capacity to improve ourselves on the basis of reason and judgment."[18] If they are right about the spirit of sport, then it seems we'd prefer watching posthuman sprinters to human ones. Superhuman performances are more likely to come from the former than the latter. Those who aspire to be top-level athletes should be availing themselves of any safe, competitively useful GNR technology.

In what follows, I argue that our interest in elite sport is governed by a value that opposes enhancement. This value explains why we prefer the inferior performances of human athletes to the superior ones of post-humans. Performance-enhancing drugs and modifications alienate us from our sporting heroes.

The Interest in Identifying

There does seem to be some support for the view of sport taken by Savulescu and his coauthors. Consider the confession of the *Financial Times* journalist David Owen about his experience as a spectator of the single most famous instance of drug cheating. Owen says "I have a guilty secret. I think Ben Johnson's 'victory' in the men's 100m at the 1988 Seoul Olympics is just about the most exciting 10 seconds of sport I have ever witnessed."[19] If the number of clips of Johnson's race available on YouTube is any indication then Owen may not be his only secret admirer.

We can explain the excitement that Johnson's performance elicits by appealing to a particular interest that we have in sport. Call this an *interest in extremes*. The people of Victorian England were spurred by this interest to attend exhibitions of the world's tallest, shortest, strongest, and most wolf-like humans. The 9.84-second time recorded by Johnson at Seoul was fascinating because it was a behavioral extreme—at that time the fastest 100-meter dash in history. Athletic performances are not the only things that engage our interest in extremes. It explains the popularity of Discovery Channel documentaries on the biggest dinosaurs and the highest mountains.

I think, though, that the interest in extremes is a relatively insignificant contributor to our overall enjoyment of sport. It does not explain the durability of our commitment to elite sport.

Consider the kind of behavior typically prompted by the interest in extremes. How often do you pay to see an exhibition featuring the world's tallest human being? You may go once to satisfy your curiosity. But I suspect that you're unlikely to go more often than that. Once you've been into the tent with the world's tallest human being, how likely are you to choose to pay a dollar to enter the tent with the world's tenth-tallest human being? I suspect not particularly. This pattern is unlike that of sports fans, who tune in and turn up on a weekly or even daily basis. This week's sporting performances can give pleasure even if we know that they're objectively less impressive than last week's.

I'm confident that games between human players will continue to interest us to a much greater extent than those between chess supercomputers, or those between radically enhanced posthumans. We continue to watch human 1,500-meter runners when we know that any used car lot contains many machines that cover the distance more quickly. Garry Kasparov is just more interesting to us than the chess computers Deep Blue, Deep Junior, or any future machine that inherits the "Deep" mantle. He'd be more interesting to us than any posthuman ten-dimensional chess players slumming it by playing two-dimensional chess.

So what could the basis of this preference for the inferior performances of humans be? Is it rationally defensible?

In what follows I describe a distinct interest that we have in elite sport. This is an *interest in identifying*. It corresponds with a human value that places limits on how good the performances of elite athletes should be. They should be sufficiently good that we enjoy watching them. But they should not be so good that they are well beyond what beings like us can do.

Vicariously Participating in Exceptional Performances

I will argue that unenhanced human spectators are drawn to the achievements of unenhanced human competitors because we recognize them as pushing up against the limits of our own activities. Their performances tell us something about what we could achieve, or might have achieved had we been more fortunate, dedicated, and talented. We recognize ourselves as sharing many of the limitations of human competitors, and this gives us a unique access to and interest in their exceptional achievements.

Differently empowered beings lack both this distinctive first-person access to and enjoyment of exceptional human performances. They are, as a consequence, unlikely to care about how close to two hours human marathoners are capable of running. We have a correspondingly diminished interest in their achievements. Aliens and posthumans may engage in sporting competitions that are exciting to them and engage their particular values. But these competitions don't have the same allure for us, even if we recognize that they are objectively exceptional. We would rather watch Usain Bolt run 100 meters in 9.58 seconds than see athletically superior posthumans or aliens run that distance much more quickly.

The terminology of a prominent theory of the human mind offers a convenient way to describe the nature of the access that unfit, unmotivated human beings have to the achievements of elite athletes, and the enjoyment we get out of viewing them.[20] Simulation theory addresses the problem of how we predict and explain the behavior of other human beings.[21] It holds that we represent the mental processes that cause the behavior of other humans by simulating them. The results of this simulation enable us to predict what they will do, and to explain why they did what they did. For example, suppose we see a hungry person poised before a large bowl of spaghetti bolognese. We imagine ourselves hungry in those circumstances. The output of our simulation is likely to be a desire to eat. Simulation theory's great advantage is that it solves the problem of how we predict and explain the behavior of other human beings without requiring us to know a massive collection of facts about them. We use our own psychological processes to simulate theirs. When we do this we take this mental machinery "off-line" so that it produces predictions and explanations of others rather than causing us to act in certain ways. In simulating the mental processes of the hungry person we understand that *she* desires to eat the pasta; the output of the simulation isn't a desire that *we* eat.

Simulation theory assumes sufficient similarity between the individual doing the simulating and the individual being simulated. If the mental processes of intelligent aliens, posthumans, or computers are significantly different from our own, then our simulations will not be of much use in predicting or explaining their behavior.

Gregory Currie has appealed to simulation theory to explain the experience of reading fiction.[22] He claims that when we read a book, we simulate: We engage emotionally with the pleasures and pains of fictional characters by simulating their mental states. Currie says, "[i]f our imagining goes well, it will tell us something about how we would respond to the situation, and what it would be like to experience it."[23]

I think simulation theory offers a plausible description of our engagement with sport and, moreover, that fans should recognize it as capturing much of the phenomenology of the experience of watching sport. Attentive readers of Tolstoy's *Anna Karenina* vicariously experience the central character's loves and disappointments. We get something similar out of watching sport. It's occasionally useful to be able to predict the behavior of someone seated in front of a bowl of spaghetti. But doing so is not especially exciting—unless perhaps the eater is involved in a spaghetti-eating competition of the kind that used to be reported in the *Guinness Book of Records*. Simulating the mental processes behind great performances enables us to sample the exhilaration experienced by the performers. We identify with elite athletes and vicariously participate in their triumphs—and disappointments. When we simulate a great soccer goal, an incredible tennis passing shot, a daring queen sacrifice in chess, or an amazing boxing combination, we sample the exhilaration that accompanies them. We watch elite sports to vicariously experience exceptional performances and to learn something about what is possible for human beings.

Consider two descriptions of the pivotal event in the 1986 FIFA World Cup soccer quarter-final match between Argentina and England.

(A) Diego Maradona scored a goal in the 56th minute.

(B) "Fed by Hector Enrique, Maradona turned through 180 degrees out on the right, on the halfway line, before slipping between Peter Reid and Peter Beardsley. Next he sped inside centre-back Terry Butcher and fended off a challenge from Terry Fenwick, who had been distracted by the lurking presence of the advancing Jorge Valdano. Maradona slalomed on deep into the penalty box, waited for Peter Shilton to step from his line then dummied left before stepping right to slip the ball past the keeper and over the line just as the recovering Butcher launched another, vain tackle."[24]

I suspect that description (B) makes much more exciting reading because it engages our simulation machinery. It guides us through the movements performed by Maradona as he worked the ball though half the England team, and it encourages us to imagine doing the same. We get to feel something of the exhilaration of Maradona when he completes the movement. Masochists may instead choose to simulate the mental states of the English defenders.

Our identification with the participants in elite sport is premised on their being sufficiently similar to us for their performances to be relevant to us at the same time as their being sufficiently different for their performances to be exceptional. The vicarious participation enabled by simulation is possible because of some characteristics that top human performers share with the rest of us. To paraphrase the 1992 Gatorade commercial featuring the basketball legend Michael Jordan, for us to want to try to "be like Mike," Mike has to be sufficiently like us.

Anyone who's attempted to jog up and down his or her street has some insight into the experiences of Paula Radcliffe when she was producing her record-breaking performances. Though Garry Kasparov may be the best human chess player there has ever been, even a thoroughly mediocre player can move the chess pieces as he and Anatoly Karpov moved them in the celebrated sixteenth game of their 1985 match. We can read Kasparov's annotations and thereby sample the pleasure he experienced in defeating the then world champion.

We humans want to watch athletes and performers who are human; elite athletes should want to produce performances that are relevant to their human audiences. I think this is the basis of our complaint about drug cheats. The performances of human drug cheats are more relevant to us than are the performances of posthuman competitors. But they're less relevant than are the performances of clean human competitors. The triumphs and failures of posthuman sportspeople may interest the statisticians among us, but they're unlikely ever to pack the emotional interest of the final one hundred meters of a Paula Radcliffe marathon or of a well-taken Rafael Nadal forehand.

What should defenders of radical enhancement take from the preceding species-relativist discussions? It's possible that they are genuinely alienated from their own humanity and so fail to see any value in the experiences I've described. Their engagement with elite sport may be guided only by the interest in extremes. But if this is so, I'm guessing that they're in the minority. The rest of us want to see human athletes producing performances relevant to human audiences.

The Choice

When Julian Huxley first issued his call for radical enhancement, the GNR technologies that feature strongly in the plans of Kurzweil, de Grey,

Bostrom, and Hughes were either nonexistent or in a primitive state. Huxley's proposal that we assume a managing directorship over the business of human evolution required no futuristic technology. He was an advocate of eugenics.[25] A eugenics program aiming at radically enhancing humans would both encourage the relatively long-lived and intelligent among us to have many children and discourage the relatively short-lived and unintelligent from having any children. If both of the positive and negative parts of this program were successfully implemented, then it is possible, over many generations, that there would be increases in intelligence and extensions of life spans.

Huxley didn't use the word "eugenics" in his 1957 paper. His reticence on this score is not hard to understand. Huxley's readers had clear memories of the tragic eugenic policies of the Nazis. Whereas we are accustomed to thinking of the Holocaust as a singular moral catastrophe, it can be viewed as the most brutal manifestation of Nazi negative eugenics. For the Nazis, murder was the most direct and decisive way of stopping people they despised from having children.

Contemporary transhumanism's mechanism of radical enhancement is, of course, completely different from that of Huxley's eugenics program. The technologies that are the focus of this book may end up redirecting human evolution, but they will not do this by manipulating evolutionary forces. People will be free to make their own choices about whether and how to radically enhance themselves or their children. Contemporary radical enhancers certainly aren't advocating murder or sterilization or any of the other methods eugenicists hoped to use to redirect human evolution.

We're now in a position to identify a problem for this technological route to radical enhancement. Huxley's evolutionary path would see humans traveling the path to a radically enhanced future collectively. It would also take time. Over many generations, it could lead to beings that are so different from us that they are not properly considered human. But the gradual nature of this transformation wouldn't prevent us from relating to our former selves, our children, and our fellow citizens in ways compatible with our human values. We and they will be human. The technological transformations advocated by early twenty-first century defenders of radical enhancement differ in being abrupt. They threaten the bonds of kinship and fellow feeling that connect us with other humans, taking much of the meaning out of human triumphs and tragedies.

The opposition to radical enhancement defended in this book does not depend on the absurd conceit that the human species will last forever. Eventually, should some global or galactic catastrophe not leave us completely without descendants, we'll evolve into beings who are not properly considered human. Posthumanity certainly coincides more closely with our human values than do H. G. Wells's Eloi or Morlock possible futures for the human species. So it may be reasonable to prefer a posthuman distant future to either of the Wellsian alternatives. But that nonhuman future should ideally be a distant one. Our human values direct us to retain our humanity over the next few centuries in preference to becoming Eloi, Morlock, posthuman, or cyborg.

We should approach the seeming inevitability of our species' end much as we confront the prospect of our own personal demise. We know, *pace* de Grey, that our lives will end. When we die, many aspects of our lived experience will be lost forever. The idiosyncratic combination of qualities, defects, and experiences that defines us will go out of existence, leaving to our loved ones only comparatively brief narrative outlines of our lives. The tragedy of death sounds like a reason to sign up for SENS. But I've argued that many of the things that we care about do not survive the radical extension of our life spans. We may persist, but only with existences that we properly view as impoverished. But equally, the inevitability of death is no reason to commit suicide. So it is with the idiosyncratic combination of qualities, defects, and experiences that we recognize as making us and others human. The human species won't last forever, but that's no reason to either expedite its end or to remove ourselves from it. We should instead enjoy it while it lasts.

Notes

Acknowledgments

1. Nicholas Agar, "Whereto Transhumanism? The Literature Reaches a Critical Mass," *Hastings Center Report* 37, no. 3 (2007): 12–17. For responses by James Hughes and Nick Bostrom and my reply, see "Human vs. Posthuman," *Hastings Center Report* 37, no. 5 (2007): 4–6.

1   What Is Radical Enhancement?

1. A useful starting point for those interested in transhumanism is "The Transhumanist FAQ," collated by Nick Bostrom and available at http://humanityplus.org/learn/philosophy/faq/. Among the most visible transhumanists are Nick Bostrom, James Hughes, Max More, Natasha Vita More, and Mark Walker. Transhumanism seems to exhibit the same splittist tendencies as the early twentieth-century revolutionary movements. Transhumanists certainly don't speak with a single voice. This book explores a selection of Bostrom's and Hughes's views. Those interested in a broader survey of transhumanist opinion should see Nick Bostrom's history "A History of Transhumanist Thought," *Journal of Evolution and Technology* 14, no. 1 (2005) (available on Bostrom's homepage http://www.nickbostrom.com/); James Hughes's survey of the movement *Citizen Cyborg: Why Democratic Societies Must Respond to the Redesigned Human of the Future* (Cambridge, Mass.: Westview, 2004); various items on Max More's homepage (http://www.maxmore.com/); Natasha Vita More's artwork, particularly her "future body prototype," primo posthuman (http://www.natasha.cc/primo.htm); and the many papers on Mark Walker's Web site (http://ieet.org/index.php/IEET/bio/walker/) especially "Prolegomena to Any Future Philosophy," "Genetic Virtue," "In Praise of Bio-happiness," "A neo-Irenean Theodicy," and "Ship of Fools."

2. Julian Huxley, "Transhumanism," in *New Bottles for New Wine* (London: Chatto & Windus, 1957).

3. Ibid., 13.

4. Ibid., 16.

5. Nick Bostrom, "Why I Want to Be a Posthuman When I Grow Up" in *Medical Enhancement and Posthumanity*, ed. Bert Gordijn and Ruth Chadwick (Dordrecht: Springer, 2009), 108. See also the "What Is a Posthuman?" entry on the Transhumanist FAQ (http://www.transhumanism.org/index.php/WTA/faq21/56/).

6. Bostrom, "Why I Want to Be a Posthuman When I Grow Up," 108–109.

7. Ray Kurzweil's presentations of radical enhancement include, but are not limited to *The Age of Intelligent Machines* (Cambridge, Mass.: MIT Press, 1990), *The Age of Spiritual Machines: When Computers Exceed Human Intelligence* (London: Penguin, 2000), and *The Singularity Is Near: When Humans Transcend Biology* (London: Penguin, 2005). Kurzweil's Web site is also essential viewing: http://www.kurzweilai.net/. Aubrey de Grey's book, *Ending Aging: The Rejuvenation Breakthroughs That Could Reverse Human Aging in Our Lifetime* (New York: St Martin's Press, 2007), written with Michael Rae, presents the science behind extended life spans. A large collection of academic and popular papers can be found on de Grey's Web site: http://www.mfoundation.org/. On Nick Bostrom's well-laid-out Web site, http://www.nickbostrom.com/, publications are signposted in terms of his own views about their overall importance and their intended audiences. James Hughes presents his take on the social context and consequences of radical enhancement in *Citizen Cyborg*. Recordings from Hughes's radio show *Changesurfer Radio* can be found at http://ieet.org/index.php/IEET/csr.

8. Kurzweil, *The Singularity Is Near*, 7.

9. Ibid., 136.

10. Francis Fukuyama, *Our Posthuman Future: Consequences of the Biotechnology Revolution* (New York: Farrar, Straus & Giroux, 2002); Leon Kass, "The Wisdom of Repugnance: Why We Should Ban the Cloning of Humans", *New Republic*, June 2, 1997; Leon Kass, *Life, Liberty, and the Defense of Dignity: The Challenge for Bioethics* (San Francisco: Encounter Books, 2002); Bill McKibben, *Enough: Staying Human in an Engineered Age* (New York: Times Books, 2003).

11. Nick Bostrom and Toby Ord, "The Reversal Test: Eliminating Status Quo Bias in Applied Ethics," *Ethics* 116 (2006): 656–679.

12. For philosophical defenses of cultural relativism, see Gilbert Harman, "Moral Relativism," in *Moral Relativism and Moral Objectivity*, ed. G. Harman and J. J. Thompson (Cambridge, Mass.: Blackwell, 1996); and David Wong, *Moral Relativity* (Berkeley: University of California Press, 1984).

13. See Steven Pinker, *The Blank Slate: The Modern Denial of Human Nature* (New York: Viking Books, 2002) for a prominent recent defense of the influence of nature on human psychology.

14. See Peter Singer, *Practical Ethics* (Cambridge: Cambridge University Press, 1993), and *Animal Liberation: A New Ethics for our Treatment of Animals* (New York: Random House, 1975).

2 Radical Enhancement and Posthumanity

1. This distinction leaves open the additional category of slightly less than radical enhancement, which would include enhancements beyond human extremes but not greatly so.

2. For prominent recent arguments that parents who moderately genetically enhance their offspring exercise illegitimate control, see Michael Sandel, *The Case against Perfection: Ethics in the Age of Genetic Engineering* (Cambridge, Mass.: Harvard University Press, 2009), and Jürgen Habermas, *The Future of Human Nature* (Cambridge: Polity Press, 2003).

3. Julian Huxley, "Transhumanism," in *New Bottles for New Wine* (London: Chatto & Windus, 1957), 16.

4. See Bostrom's "What Is a Posthuman?" entry on the Transhumanist FAQ, http://humanityplus.org/learn/philosophy/faq#answer_20, and James Hughes, *Citizen Cyborg: Why Democratic Societies Must Respond to the Redesigned Human of the Future* (Cambridge, Mass.: Westview, 2004), 97–103.

5. Taken from Kurzweil's response to a question at a 2005 Harvard University conference regarding the relationship between human and artificial intelligence. See http://en.wikipedia.org/wiki/Raymond_Kurzweil.

6. See, e.g., Nick Bostrom and Toby Ord, "The Reversal Test: Eliminating Status Quo Bias in Applied Ethics," *Ethics* 116 (2006): 656–679.

7. See Felipe Fernandez-Armesto, *So You Think You're Human? A Brief History of Humankind* (Oxford: Oxford University Press, 2004) for a recent expression of skepticism about a scientific definition of humanity.

8. The biological species concept was originally formulated and defended by the biologist Ernst Mayr. See his book *Systematics and the Origin of Species* (New York: Columbia University Press, 1942).

9. See, e.g., *Species: New Interdisciplinary Essays*, ed. Robert Wilson (Cambridge, Mass.: MIT Press, 1999).

10. See Fernandez-Armesto, *So You Think You're Human?* for a useful summary of the history of attempts to capture humanity's essential properties.

11. For a defense of the view that apes possess the "building blocks of morality," see Frans de Waal, *Good Natured: The Origins of Right and Wrong in Humans and Other Animals* (Cambridge, Mass.: Harvard University Press, 1996).

12. Francis Fukuyama, *Our Posthuman Future: Consequences of the Biotechnology Revolution* (New York: Farrar, Straus & Giroux, 2002).

13. For skepticism about moral appeals to human nature, see Allen Buchanan, "Human Nature and Enhancement," *Bioethics* 23, no. 3 (2009): 141–150.

14. Allen Buchanan, "Moral Status and Human Enhancement," *Philosophy and Public Affairs* 37, no. 4 (2009): 346–381.

15. Ibid., 351, note 10.

16. David Levy, *Love and Sex with Robots: The Evolution of Human–Robot Relationships* (New York: HarperCollins, 2007).

17. For the related point that one could undergo enhancement that made one no longer human without losing one's identity, see note 7 on p. 144 of Allen Buchanan, "Human Nature and Enhancement," *Bioethics* 23, no. 3 (2009): 141–150.

18. For a recent assertion that they did, see Gregory Cochran and Henry Harpending, *The 10,000 Year Explosion: How Civilization Accelerated Human Evolution* (New York: Basic Books, 2009).

19. Kurzweil, *The Singularity Is Near: When Humans Transcend Biology* (London: Penguin, 2005), 374.

20. Ibid., 318–320. For a more fully realized account of human–machine friendships and sexual interactions see Levy, *Love and Sex with Robots*.

21. Jim Giles, "What Puts the Creepy into Robot Crawlies?" *New Scientist* 27, no. 2627 (October 2007).

22. Fukuyama, *Our Posthuman Future*, chapter 3.

23. Donna Chisholm, "NZ's First Test-Tube Baby Turns 25," *North and South* (June 2009).

24. Ibid., p. 41.

25. Silver, *Remaking Eden: How Genetic Engineering and Cloning Will Transform the American Family* (New York: Avon Books, 1997), 5.

26. Ibid., 6.

27. Gregory Stock, *Redesigning Humans: Our Inevitable Genetic Future* (Boston: Houghton Mifflin, 2002), chapter 4.

28. http://humanityplus.org/learn/philosophy/faq#answer_20.

29. Adapted from Nick Bostrom, "Human Genetic Enhancements: A Transhumanist Perspective," *Journal of Value Inquiry* 37, no. 4 (2003): section 2.

30. See David Heyd, *Genethics: Moral Issues in the Creation of People* (Berkeley: University of California Press, 1994), chapter 8, for a discussion of the idea of vicariously surviving through our children.

3 The Technologist—Ray Kurzweil and the Law of Accelerating Returns

1. Ray Kurzweil, *The Singularity Is Near: When Humans Transcend Biology* (London: Penguin, 2005), 7. Vernor Vinge was the first to talk about a technological singularity, drawing an analogy between technological change and the effects of a gravitational singularity at the heart of a black hole. See Vernor Vinge, "The Coming Technological Singularity: How to Survive in the Post-Human Era," *Vision-21: Interdisciplinary Science and Engineering in the Era of CyberSpace* (1993), http://www-rohan.sdsu.edu/faculty/vinge/misc/singularity.html.

2. Kurzweil, *The Singularity Is Near*, 23.

3. See http://en.wikipedia.org/wiki/Raymond_Kurzweil#Accuracy_of_predictions.

4. Comment appears on the dust jacket of Kurzweil, *The Singularity Is Near*.

5. Both of these questions were answered correctly on the UK version of *Who Wants to Be a Millionaire?* A correct answer to the first question netted the contestant £1,000,000. A correct answer to the second was deemed fraud and earned its answerer a suspended jail sentence.

6. See ftp://download.intel.com/museum/Moores_Law/Video-Transcripts/Excepts_A_Conversation_with_Gordon_Moore.pdf.

7. For skepticism about Kurzweil's law of accelerating returns, see Alfred Nordmann, "Singular Simplicity," Special Report: The Singularity, *IEEE Spectrum Online*, http://www.spectrum.ieee.org/jun08/6273.

8. Kasparov did manage to avenge humanity on the machines. In 2003 he tied a six-game match against Deep Junior, the then world champion among chess computers. The view among chess experts (admittedly, human chess experts) was that Kasparov was, on balance, the better player throughout the match.

9. Quoted in Rudy Chelminski, "This Time It's Personal: Humankind Battles to Reclaim the Chess-Playing Championship of the World," *Wired 9.10*, October 2001.

10. Kurzweil, *The Singularity Is Near*, 122–125.

11. Ibid., 126–127.

12. Ibid., 125–126.

13. http://www-03.ibm.com/press/us/en/pressrelease/24405.wss.

14. Kurzweil, *The Singularity Is Near*, 197.

15. Ibid., 200.

16. Find details of Kurzweil's bet with Mitch Kapor, the founder of the Lotus Development Corporation, at http://www.longbets.org/1.

17. Kurzweil, *The Singularity Is Near*, 127.

18. Ibid., 145.

19. Duncan Graham-Rowe, "Do We Have the Technology to Build a Bionic Human?" *New Scientist* (July 4, 2008).

20. Ray Kurzweil, "The Evolution of Mind in the Twenty-First Century," in *Are We Spiritual Machines: Ray Kurzweil vs. the Critics of Strong AI*, ed. Jay Richards (Seattle: Discovery Institute Press, 2002), 14.

21. Ray Kurzweil, *The Age of Spiritual Machines: When Computers Exceed Human Intelligence* (New York: Viking 1999), 105.

22. Ibid., 126.

23. Kurzweil, *The Singularity Is Near*, 9.

24. Ibid., 29.

25. Ibid.

26. Ibid., 9.

27. John Horgan, "The Consciousness Conundrum," Special Report: The Singularity, *IEEE Spectrum Online*, http://www.spectrum.ieee.org/jun08/6280.

28. Peter D. Kramer, *Listening to Prozac* (New York: Viking, 1993), quoted in Kurzweil, *The Singularity Is Near*, 169.

29. Kurzweil, *The Singularity Is Near*, 169.

30. Ibid., 168.

31. Ibid., 145.

32. Ibid., 169.

33. Ibid., 168.

34. Roger Penrose, *Shadows of the Mind: A Search for the Missing Science of Consciousness* (Oxford: Oxford University Press, 1994).

35. Kurzweil, *The Singularity Is Near*, 450–452.

36. Consider, for example, the theory of Francis Crick and Christof Koch that 40 hertz oscillations in the visual cortex and elsewhere explain the binding of various types of information into a unified whole. This is certainly an interesting proposal.

But the synthesizing function of consciousness is just one part of the overall phenomenon. Francis Crick and Christof Koch, "Consciousness and Neuroscience," *Cerebral Cortex* 8 (1998): 97–107.

37. Kurzweil, *The Singularity Is Near*, 30.

38. Ibid., 194–195.

39. Ibid., 374.

40. Ibid., 310.

41. Ibid., 386.

4   Is Uploading Ourselves into Machines a Good Bet?

1. Susan Schneider deploys considerations about personal identity to challenge the rationality of uploading. She argues that the notion that one could upload depends on an untenable version of the psychological continuity theory of our identities. The second manner of uploading offers the best prospects of a response to Schneider. A candidate for uploading would allow time for each new electronic chip to integrate itself into his or her psychology before proceeding. See Susan Schneider, *"Mindscan*: Transcending and Enhancing the Human Brain," in *Science Fiction and Philosophy: From Time Travel to Superintelligence*, ed. Susan Schneider (Oxford: Wiley-Blackwell, 2009).

2. Searle's argument was first presented in John Searle, "Minds, Brains and Programs," *Behavioral and Brain Sciences* 3, no. 3 (1980): 417–457.

3. For Kurzweil's responses, see Ray Kurzweil, "Locked in His Chinese Room: Response to John Searle," in *Are We Spiritual Machines? Ray Kurzweil vs. the Critics of Strong A.I.*, Jay W. Richards (Seattle, Wash.: Discovery Institute, 2002).

4. Kurzweil, *The Singularity Is Near*, 380.

5. See section 4.3.1 "What is cryonics?" at http://www.extropy.org/faq.htm.

6. Ibid.

7. http://www.benbest.com/cryonics/CryoFAQ.html#_VR.

8. Kurzweil, *The Singularity Is Near*, 127.

9. Ibid., 145.

10. Ibid., 29.

11. Ibid.

12. Daniel Dennett explores a variety of puzzles such as these concerning phenomenal consciousness in his book *Consciousness Explained* (Boston: Little, Brown, 1991).

Dennett concludes that the notion of phenomenal consciousness is incoherent. His arguments may make those who reject Searle's Wager more confident about uploading. But they should not be viewed as reducing to zero either the probability that biological human brains support phenomenal consciousness or that uploading destroys this capacity.

13. Kurzweil, *The Singularity Is Near*, 378–379.

14. http://www.singinst.org/upload/CFAI.html#challenge.

15. Kurzweil, *The Singularity Is Near*, 431.

16. Ibid., 341.

17. Nick Bostrom, "Existential Risks: Analyzing Human Extinction Scenarios and Related Hazards," *Journal of Evolution and Technology* 9 (2002).

18. Ibid.

19. Jerry Fodor, *The Modularity of Mind: An Essay on Faculty Psychology* (Cambridge, Mass.: MIT Press, 1983).

5 The Therapist—Aubrey de Grey's Strategies for Engineered Negligible Senescence

1. This is the theme of Ray Kurzweil and Terry Grossman, *Fantastic Voyage: Live Long Enough to Live Forever* (New York: Plume Books, 2005).

2. Aubrey de Grey and Michael Rae, *Ending Aging: The Rejuvenation Breakthroughs That Could Reverse Human Aging in Our Lifetime* (New York: St. Martin's Press, 2007), 336.

3. For the estimate on the cost to send humans to the moon, see http://spaceflight.nasa.gov/history/apollo/index.html.

4. See de Grey's presentation "Fixing Humanity's Worst Problem" to the 2006 TED (Technology, Entertainment, Design) Conference, http://video.google.com/videoplay?docid=3847943059984264388.

5. De Grey and Rae, *Ending Aging*, 14.

6. Aubrey de Grey, "An Engineer's Approach to the Development of Real Anti-Aging Medicine," http://sageke.sciencemag.org/cgi/content/full/2003/1/vp1.

7. http://www.sens.org/index.php?pagename=sensf_faq_timeframe.

8. Aubrey de Grey, "Why We Age and How We Can Avoid It," http://www.ted.com/index.php/talks/view/id/39.

9. De Grey and Rae, *Ending Aging*, 8.

10. Aubrey de Grey, "The War on Aging," in the Immortality Institute 2004 collection, *The Scientific Conquest of Death: Essays on Infinite Lifespans* (Buenos Aires: Libros en Red, 2004), 33.

11. De Grey and Rae, *Ending Aging*, chapter 11.

12. Ibid., chapter 10.

13. Ibid., chapter 12.

14. Ibid., chapters 5 and 6.

15. Ibid., chapters 7 and 8.

16. Ibid., chapter 9.

17. Figures taken from the American Cancer Society: http://www.cancer.org/docroot/STT/content/STT_1x_Cancer_Facts_Figures_2009.asp.

18. "From Here to Eternity," *Discover* (summer 2009): 20–22.

19. Interview with H. Gilbert Welch by Andrew Grant, "The Cancer Contrarian," *Discover* (summer 2009): 24–26.

20. Ibid., 24.

21. Ibid., 26.

22. De Grey and Rae, *Ending Aging*, chapter 12.

23. Ibid., 307–308.

24. Ibid., 299.

25. Linda Geddes, "A Small Step Closer to Eternal Youth," *New Scientist* 200, no. 2683 (November 22, 2008): 8–9.

26. Ibid., 9.

27. De Grey, quoted in Geddes, "A Small Step Closer to Eternal Youth," 9.

28. De Grey and Rae, *Ending Aging*, 308.

29. De Grey, quoted at Methuselah Foundation, "Concerns," http://www.sens.org/index.php?pagename=sensf_faq_concerns.

30. De Grey and Rae, *Ending Aging*, chapter 7.

31. Aubrey de Grey, "Escape Velocity: Why the Prospect of Extreme Human Life Extension Matters Now," *Public Library of Science: Biology* 2, no. 6 (2004): 725.

32. Aubrey de Grey, "The Quest for Indefinite Life III: The Progress of SENS," http://rationalargumentator.com/issue24/indefinitelife3.html.

33. Ibid.

34. Viewable at http://www.acceleratingfuture.com/michael/blog/?p=466.

35. Preston W. Estep, III, Matt Kaeberlein, Pankaj Kapahi, Brian K. Kennedy, Gordon J. Lithgow, George M. Martin, Simon Melov, R. Wilson Powers, III, and Heidi A. Tissenbaum, "Life-Extension Pseudoscience and the SENS Plan," *MIT Technology Review* 109, no. 3 (2006): 84.

36. Jason Pontin, "Is Defeating Aging Only a Dream? No One Has Won Our $20,000 Challenge to Disprove Aubrey de Grey's Anti-aging Proposals," *Technology Review* 109 (2006): 80–84, http://www.technologyreview.com/sens/index.aspx.

6   Who Wants to Live Forever?

1. See Felicia Nimue Ackerman's very good discussion of ethical issues arising out of life extension in "Death Is a Punch in the Jaw: Life-Extension and Its Discontents," in *The Oxford Handbook of Bioethics*, ed. Bonnie Steinbock (Oxford: Oxford University Press, 2007).

2. Joel Garreau, "The Invincible Man: Aubrey de Grey, 44 Going on 1,000, Wants Out of Old Age," *Washington Post*, October 31, 2007, http://www.washingtonpost.com/wp-dyn/content/article/2007/10/30/AR2007103002222.html.

3. Mark Walker makes the interesting point that negligible senescence doesn't have to be particularly popular to spread. Its spread may owe more to the fact that those who choose it tend to significantly outlive those who do not. See Mark Walker, "Universal Superlongevity: Is It Inevitable and Is It Good?," http://ieet.org/index.php/IEET/more/walker20060217/.

4. Ibid.

5. Ibid.

6. Bryan Appleyard, *How to Live Forever or Die Trying* (New York: Simon and Schuster, 2007), 5.

7. Bernard Williams, "The Makropulos Case: Reflections on the Tedium of Immortality," in *Problems of the Self* (Cambridge: Cambridge University Press, 1973), 423.

8. Christine Overall, *Aging, Death, and Human Longevity: A Philosophical Inquiry* (Berkeley: University of California Press, 2003), 145.

9. John Harris, "A Response to Walter Glannon," *Bioethics* 16, no. 3 (2000): 284.

10. Mark Walker, "Ennui and Superlongevity," http://www.nmsu.edu/~philos/documents/boredom.doc.

11. Interview with Aubrey de Grey, *Speculist*, August 8, 2003, http://www.speculist.com/archives/000065.html.

12. Ibid.

13. Ibid.

14. See http://www.phmsa.dot.gov/hazmat/risk/library.

15. See http://www.imdb.com/title/tt0449059/quotes.

16. Aubrey de Grey, personal communication.

17. "David Deutsch Speaks with Aubrey de Grey about SENS," http://www
.acceleratingfuture.com/michael/blog/2008/05/david-deutsch-speaks-with-aubrey
-de-grey-about-sens/.

18. Ibid.

19. Ibid.

20. For a very good popular exposition of the Red Queen hypothesis, see Matt
Ridley, *The Red Queen: Sex and the Evolution of Human Nature* (New York: Harper
Perennial, 2003).

21. See http://www.rael.org/.

22. Aubrey de Grey, personal communication.

23. Weijing Hel, Stuart Neil, Hemant Kulkarnil, Edward Wright, Brian K. Agan,
Vincent C. Marconi, Matthew J. Dolan, Robin A. Weiss, and Sunil K. Ahujal, "Duffy
Antigen Receptor for Chemokines Mediates Trans-infection of HIV-1 from Red Blood
Cells to Target Cells and Affects HIV-AIDS Susceptibility," *Cell Host and Microbe* 4
(2008): 52–62.

24. Aubrey de Grey, personal communication.

25. See, e.g., Max Hastings's contrast between the attitudes toward killing and dying
of the Western Allied soldiers and those of the soldiers of Germany and of the Soviet
Union. Max Hastings, *Armageddon: The Battle for Germany, 1944–1945* (London:
Vintage Books, 2005).

26. To sample the controversy, see Alex Berenson, "Cancer Drugs Offer Hope, But
at a Huge Expense," *New York Times*, July 12, 2005, http://query.nytimes.com/gst/
fullpage.html?res=9503E5DC133DF931A25754C0A9639C8B63&sec=health.

27. See Leonard Fleck, "The Costs of Caring: Who Pays? Who Profits? Who Panders?"
Hastings Centre Report 36, no. 3 (2006).

28. Aubrey de Grey and Michael Rae, *Ending Aging: The Rejuvenation Breakthroughs
That Could Reverse Human Aging in our Lifetime* (New York: St. Martin's Press, 2007),
312.

29. Ibid., chapter 13.

30. Among the most morally repellent examples of Nazi medical research were the hypothermia experiments. For information on these, see the materials at http://www.pbs.org/wgbh/nova/holocaust/experiside.html. For skepticism about their scientific value, see Lawrence Altman, "Nazi Data on Hypothermia Termed Unscientific," *New York Times*, May 17, 1990, http://query.nytimes.com/gst/fullpage.html?res=9C0CE4D7163BF934A25756C0A966958260.

31. De Grey actually argues that we should significantly relax rules regarding participation in medical experiments. See the discussion in de Grey and Rae, *Ending Aging*, 323–324.

7   The Philosopher—Nick Bostrom on the Morality of Enhancement

1. Nick Bostrom and Toby Ord, "The Reversal Test: Eliminating Status Quo Bias in Applied Ethics," *Ethics* 116 (2006): 656–679.

2. Thomas Gilovich, Dale Griffin, and Daniel Kahneman, *Heuristics and Biases: The Psychology of Intuitive Judgment* (Cambridge: Cambridge University Press, 2002).

3. Bostrom and Ord, "The Reversal Test," 658.

4. Ibid., 660. Original in Raymond S. Hartman, Michael J. Doane, and Chi-Keung Woo, "Consumer Rationality and the Status Quo," *Quarterly Journal of Economics* 106 (1991): 141–162.

5. Bostrom and Ord, "The Reversal Test," 664–665.

6. Ibid., 672.

7. Ibid., 669.

8. Ibid..

9. Nick Bostrom, "Human Genetic Enhancements: A Transhumanist Perspective," *Journal of Value Inquiry* 37 (2003): 493–506, at 495.

10. David Lewis, "Dispositional Theories of Value," *Proceedings of the Aristotelian Society* suppl. vol. 63 (1989): 113–137.

11. Bostrom, "Human Genetic Enhancements," 495.

12. Nick Bostrom, "Transhumanist Values," in *Ethical Issues for the 21st Century*, ed. Frederick Adams (Bowling Green, Ohio: Philosophical Documentation Center Press, 2003).

13. Nick Bostrom, "Why I Want to Be a Posthuman When I Grow Up," in *Medical Enhancement and Posthumanity*, ed. Bert Gordijn and Ruth Chadwick (Dordrecht: Springer, 2009), 132.

14. Nick Bostrom, letter to the *Hastings Center Report*, "Human vs. Posthuman," *Hastings Center Report*, September–October 2007.

15. Nick Bostrom, personal communication.

16. Ibid.

17. Timothy D. Wilson, Thalia Wheatley, Jonathan M. Meyers, Daniel T. Gilbert, and Danny Axsom, "Focalism: A Source of Durability Bias in Affective Forecasting," *Journal of Personality and Social Psychology* 78, no. 5 (May 2000): 821–836. Focalism is one of the topics of Daniel Gilbert's book *Stumbling on Happiness* (New York: Vintage Books, 2007).

18. Wilson et al., "Focalism," 822.

19. Ibid., 821.

20. Ibid., 835.

21. N. Liberman and Y. Trope, "The Role of Feasibility and Desirability Considerations in Near and Distant Future Decisions: A Test of Temporal Construal Theory," *Journal of Personality and Social Psychology* 75 (1998): 5–18.

22. Bostrom, "Why I Want to Be a Posthuman When I Grow Up," 111–112.

23. Ibid., 112.

8 The Sociologist—James Hughes and the Many Paths of Moral Enhancement

1. Francis Fukuyama, *The End of History and the Last Man* (New York: The Free Press, 1992).

2. Francis Fukuyama, *Our Posthuman Future: Consequences of the Biotechnology Revolution* (New York: Farrar, Straus & Giroux, 2002).

3. Thomas Jefferson, letter to Roger Weightman, Writings 1517, quoted in Fukuyama, *Our Posthuman Future*, 9.

4. Fukuyama, *Our Posthuman Future*, 10.

5. G. Annas, L. Andrews, and R. Isasi. "Protecting the Endangered Human: Toward an International Treaty Prohibiting Cloning and Inheritable Alterations," *American Journal of Law and Medicine* 28 (2002): 162.

6. James Hughes, *Citizen Cyborg: Why Democratic Societies Must Respond to the Redesigned Human of the Future* (Cambridge, Mass.: Westview, 2004), 217.

7. Allen Buchanan, "Moral Status and Human Enhancement," *Philosophy and Public Affairs* 37, no. 4: 346–381, explores the possibility that enhancement might not only replace humans with posthumans, but also bring postpersons into existence. He is

pessimistic about the prospect of explaining how enhanced postpersons could possess a moral status higher than that of persons.

8. James Wilson, "Transhumanism and Moral Equality," *Bioethics* 21 (2007): 419–425, uses a concept of moral personhood that he takes from John Rawls to support the claim that cognitive enhancement poses no threat to the moral standing of unenhanced humans. Wilson presents personhood as a *threshold property*. He explains that threshold properties are those that "do vary by degree, but where the differences in properties are significant only until a certain threshold has been reached" (422). If posthuman philosophers are as impressed by Rawls as is Wilson, then additional increments of intelligence make no difference to their concept of moral personhood. Humans and posthumans may differ in many ways, but they are both full and equal moral persons. Wilson's interesting points about personhood and moral standing don't deflect prudential concerns about the dominant moral codes of human–posthuman societies. See also Allen Buchanan, "Moral Status and Human Enhancement."

9. Hughes, *Citizen Cyborg*, 223.

10. The notion was introduced in Brin's Uplift trilogy, which includes *Sundiver* (New York: Bantam, 1980); *Startide Rising* (New York: Bantam, 1983); and *The Uplift War* (Michigan: Phantasia Press, 1987).

11. Hughes, *Citizen Cyborg*, 226.

12. Ibid.

13. Ibid., 248.

14. Ibid., 250.

15. Ibid., 253.

16. Ibid., 256.

17. Ibid., 255.

18. See Michael Smith, *The Moral Problem* (Oxford: Blackwell, 1994), for a prominent account of the relationship between moral truths and psychological dispositions.

19. Gerd Gigerenzer, *Gut Feelings: The Intelligence of the Unconscious* (New York: Viking, 2007).

20. Ibid., 85.

21. John Rawls, *A Theory of Justice* (Cambridge, Mass.: Harvard University Press, 1971); David Gauthier, *Morals by Agreement* (Oxford: Oxford University Press, 1986).

22. Rawls, *A Theory of Justice*.

23. See, e.g., Peter Carruthers, *The Animals Issue: Moral Theory in Practice* (Cambridge: Cambridge University Press, 1992). George Dvorsky argues that animals belong behind Rawls's veil of ignorance in his "All Together Now: Developmental and Ethical Considerations for Biologically Uplifting Nonhuman Animals," *Journal of Evolution and Technology* 18 (2008): 129–142, http://jetpress.org/v18/dvorsky.htm.

24. For one version of this consequentialist argument, see Aubrey de Grey, "Three Self-evident Life-extension Truths," *Rejuvenation Research* 7, no. 3 (2004): 165–167.

25. See Peter Singer, *Practical Ethics* (Cambridge: Cambridge University Press, 1993), and *Animal Liberation: A New Ethics for Our Treatment of Animals* (New York: Random House, 1975).

26. Singer, *Practical Ethics*, 105–109.

27. Ibid., 106.

28. Ibid., 107.

29. "Monkeys, Rats, and Me: Animal Testing," BBC2, first broadcast Monday, November 17, 2006.

30. See, e.g., Gareth Walsh, "Father of Animal Activism Backs Monkey Testing," *Sunday Times*, November 26, 2006, http://www.timesonline.co.uk/tol/news/uk/article650168.ece.

31. Letter to the editor of the *Sunday Times*, http://www.utilitarian.net/singer/by/20061203.htm. For further discussion, see the FAQ linked from Singer's homepage, http://www.princeton.edu/~psinger/faq.html.

32. Letter to the editor of the *Sunday Times*.

33. Ibid.

34. For accounts that don't skimp on details about the canine death toll in major medical advances, see G. Wayne Miller, *King of Hearts: The True Story of the Maverick Who Pioneered Open Heart Surgery* (New York: Three Rivers Press, 2000), and Michael Bliss, *The Discovery of Insulin: Twenty-fifth Anniversary Edition* (Chicago: University of Chicago Press, 2007).

35. James Hughes, personal communication.

36. Silver, *Remaking Eden: How Genetic Engineering and Cloning Will Transform the American Family* (New York: Avon Books, 1997), 6.

37. Hughes, *Citizen Cyborg*, xv.

38. Ibid., 253.

39. See, e.g., Kerry Lynn Macintosh, *Illegal Beings: Human Clones and the Law* (New York: Cambridge University Press, 2005).

9   A Species-Relativist Conclusion about Radical Enhancement

1. Francis Fukuyama, *Our Posthuman Future: Consequences of the Biotechnology Revolution* (New York: Farrar, Straus & Giroux, 2002), 173.

2. Francis Fukuyama, "Transhumanism," contribution in the symposium The World's Most Dangerous Ideas, *Foreign Policy* (September–October 2004).

3. McKibben, *Enough: Staying Human in an Engineered Age* (New York: Holt Paperbacks, 2004).

4. Ibid., 133.

5. Colin McGinn, "Machine Dreams," review of Francis Fukuyama, *Our Posthuman Future*, *New York Times*, May 5, 2002.

6. De Grey and Rae, *Ending Aging: The Rejuvenation Breakthroughs That Could Reverse Human Aging in Our Lifetime* (New York: St Martin's Press, 2007), 45.

7. Walter Glannon argues that life extension in itself undermines personal identity. See Walter Glannon, "Identity, Prudential Concern, and Extended Lives," *Bioethics* 16, no. 3 (2002): 266–283.

8. Nick Bostrom and Toby Ord, "The Reversal Test: Eliminating Status Quo Bias in Applied Ethics," *Ethics* 116, no. 4 (2006): 671.

9. Ibid.

10. Nick Bostrom, "Why I Want to Be a Posthuman When I Grow Up," in *Medical Enhancement and Posthumanity*, ed. Bert Gordijn and Ruth Chadwick (Dordrecht: Springer, 2009), 126–127.

11. Robertson, *Children of Choice: Freedom and the New Reproductive Technologies* (Princeton: Princeton University Press, 1994), 16.

12. Gregory Stock, *Redesigning Humans: Our Inevitable Genetic Future* (Boston: Houghton Mifflin, 2002).

13. Ronald Bailey, *Liberation Biology: The Scientific and Moral Case for the Biotech Revolution* (Amherst: Prometheus Books, 2005).

14. For my defense of the liberal view, see Nicholas Agar, *Liberal Eugenics: In Defense of Human Enhancement* (Oxford: Blackwell, 2004).

15. Dean Hamer, *The God Gene: How Faith Is Hardwired into Our Genes* (New York: Anchor Books, 2005).

16. J. Savulescu, B. Foddy, and M. Clayton, "Why We Should Allow Performance Enhancing Drugs in Sport," *British Journal of Sports Medicine* 38 (2004): 666.

17. See WADA's code at http://www.wada-ama.org/rtecontent/document/code_v2009_En.pdf.

18. Savulescu, Foddy, and Clayton, "Why We Should Allow Performance Enhancing Drugs," 667.

19. David Owen, "Chemically Enhanced," *Financial Times*, February 11, 2006.

20. Simulation theory affords a particularly intuitive presentation of our interest in identifying. But the account of our enjoyment of sport given here can be made independently of simulation theory. We learn something about the achievement of elite performers, or identify with them, by introspecting on some of our own psychological states and dispositions.

21. For a useful introduction to simulation theory, see Robert Gordon, "Folk Psychology as Mental Simulation," in *The Stanford Encyclopedia of Philosophy*, http://plato.stanford.edu/entries/folkpsych-simulation/.

22. See, e.g., Gregory Currie and Ian Ravenscroft, *Recreative Minds: Imagination in Philosophy and Psychology* (New York: Oxford University Press, 2003), and Gregory Currie, "The Moral Psychology of Fiction" in *Art and Its Messages: Meaning, Morality, and Society*, ed. Stephen Davies (University Park: University of Pennsylvania State Press, 1997).

23. Currie, "The Moral Psychology of Fiction," 56.

24. "The Greatest Goals of All Time," Telegraph.co.uk, July 4, 2007, http://www.telegraph.co.uk/sport/main.jhtml?xml=/sport/2007/07/16/nosplit/urgreatestgoals.xml.

25. See Julian Huxley, *Man in the Modern World* (London: Chatto & Windus, 1947). For histories of eugenics, see Daniel Kevles, *In the Name of Eugenics: Genetics and the Uses of Human Heredity* (Cambridge, Mass.: Harvard University Press, 1998), and Diane B. Paul, *Controlling Human Heredity: 1865 to the Present* (Atlantic Highlands, N.J.: Humanity Books, 1995).

Index

Printed in the United States
by Baker & Taylor Publisher Services